Madame Wu Chien-Shiung

The First Lady of Physics Research

Madame Wu Chien-Shiung
The First Lady of Physics Research

Chiang Tsai-Chien

Translated by
Wong Tang-Fong

World Scientific

NEW JERSEY · LONDON · SINGAPORE · BEIJING · SHANGHAI · HONG KONG · TAIPEI · CHENNAI

Published by

World Scientific Publishing Co. Pte. Ltd.

5 Toh Tuck Link, Singapore 596224

USA office: 27 Warren Street, Suite 401-402, Hackensack, NJ 07601

UK office: 57 Shelton Street, Covent Garden, London WC2H 9HE

British Library Cataloguing-in-Publication Data
A catalogue record for this book is available from the British Library.

MADAME WU CHIEN-SHIUNG
The First Lady of Physics Research

ISBN 978-981-4374-84-2
ISBN 978-981-4368-92-6 (pbk)

Printed in Singapore

Contents

*P*reface

It has been eight years since the idea of writing this biography first occurred to me. I actually started writing two years afterward. During this period, I talked to many people about Wu Chien-Shiung: some knew her a little, most had only a vague understanding; some asked me why write a biography of Wu, and some even asked who's Wu Chien-Shiung!

Who is Wu Chien-Shiung? Would we ask who is Madame Curie? Some people may ask: shall we discuss Wu Chien-Shiung and Madame Curie in the same class?

Actually, people may know that Wu Chien-Shiung was nicknamed "The Chinese Madame Curie" in the early days. In my interviews preparing for this book, many great scientists, including many foreign Nobel Laureates, said that Wu may have made greater contributions to physics than Madame Curie.

Of course, it may not be fair to compare two scientists who are half a century apart, and with very different scientific knowledge and research environments. From the point of view of a Chinese, writing a biography of Wu Chien-Shiung and honestly recording the life of a recent world-class Chinese scientist, is definitely an effort worth making.

Wu Chien-Shiung became world famous in the early 1950s for her work in nuclear physics. China was deep in civil war.

In 1956, Wu was the first to perform a rather difficult and precise experiment, and confirmed the hypothesis proposed by C. N. Yang and T. D. Lee. Yang and Lee became the first two Chinese Nobel Laureates. Although Wu did not share the prize, to the surprise and indignation of many people, she was acknowledged as one of the most distinguished experimental physicists in the world.

In 1962, Wu returned to Taiwan to attend the Congress of Academia Sinica and visited her teacher Hu Shih, but sadly witnessed the sudden death of her most beloved teacher. Wu returned again in 1965 to receive the Special Contribution Award from the Chia Hsin Cultural Foundation. She presented public lectures during both visits that contributed greatly to science in Taiwan.

As her research schedule became busier (she did not visit Taiwan again until 1983), and politics across the Taiwan Strait exerted more negative impact on sciences, her eventual homecoming to mainland China in 1973 (after an absence of 37 years) could not escape politics and criticism. In fact, this homecoming trip with both parents already passed away since her venturing abroad, and a rapidly decreasing family, had saddened her tremendously.

Wu visited Taiwan upon the invitation of Academia Sinica in 1983 after 18 years of absence. I met her for the first time as a science reporter for *Reading Times*, this meeting served as the beginning of the adventure of writing this biography.

Wu Chien-Shiung went to the US in 1936. By the time she earned a Ph.D. in 1940, her achievements and insights in research had already received the highest admiration of many professors at UC Berkeley, such as the great American scientists Oppenheimer and Lawrence. As a result, she was invited as a non-citizen to participate in the top-secret "Manhattan Project" working on atomic bombs, and made critical contribution to the project.

In a way, the participation in defense research was a chanced opportunity. Wu Chien-Shiung had worked in nuclear physics research all her life. In this field, everyone agrees that she had made at least three major influential achievements. In addition to the confirmation of the hypothesis put forth by Yang and Lee, the other two are significant enough to be considered for a Nobel Prize. In spite of her outstanding scientific achievements, she was not as well-known accordingly, perhaps because she never won a Nobel Prize.

Although Wu Chien-Shiung never won a Nobel Prize, she did receive numerous honors. The awards, medals, and honorary doctorates she received from institutes and universities form a long list, with the "Wolf Prize" endowed by an Israel industrialist probably the most representative. One criterion of the Wolf Prize is to honor candidates who deserve

Nobel Prizes but do not get one. Therefore the Wolf Prize is known as the Israeli Nobel. Wu Chien-Shiung was indeed awarded the first Wolf Prize in Physics in 1978.

In addition, she received the Cyrus B. Comstock Award (given once every five years) from the US National Academy of Sciences, National Medal of Science, and the record breaking, first female Honorary Doctorate from Princeton University in 1958. In 1975, Wu broke the white-male-president tradition, and became the first female President of the American Physical Society.

Due to her significant achievements, and her profound influence in physics, Wu Chien-Shiung was often called "The Chinese Madame Curie", "The First Lady of Physics Research", "Queen of Nuclear Research", and "The Most Distinguished Female Experimental Physicist in the World".

Wu retired from Columbia University after 36 years of teaching in 1980. As Professor Emeritus, her major research activities gradually slowed down. She was in her 70s, a good time for reflection. That's when my idea of writing this biography began.

I received much encouragement when planning for this biography, particularly from C. N. Yang, who had a special in-depth understanding of the scientific achievements of Wu Chien-Shiung, and once had a close collaboration with her.

Yang researched and taught in the US for many years and was well-known for his nurturing and directing of junior students, especially the Chinese ones. That was my impression when I first visited him in New York in 1985. When I proposed my plan to write this biography in 1987, Yang was greatly interested and warmly endorsed this effort.

Yang mentioned that a distinguished scientist like Wu Chien-Shiung deserves to have a good biography. A Chinese author, with a different perspective, would better serve this task. He suggested many ideas, and remarked about the exciting publications of many biographies of scientists in recent years in the US.

What impressed me most was his assessment of an objective narrative in these new publications. It is in contrast with many Chinese biographies, which either glorify or demote the characters, or use a novel-like, subjective narrative. Mr. Yang even recommended and accompanied me to buy a newly published book, discussing the scientists and the surrounding events leading to the progress in physics in the late 20th century.

In that exceptionally bright, sunny afternoon, I remember that Yang and I were waiting in the Stony Brook train station (it did not have a platform then), contemplating and searching for ideas. I then boarded the train back to New York City. The new book recounts the sequence of events leading to the discovery of the "J" particle by Samuel C. C. Ting. I was full of excitement and greatly moved beyond any description.

I discussed this proposal with Wu Chien-Shiung two year later. Wu was always matter-of-fact, modest, and never seeking fame, and she declined repeatedly. Only after many verbal and written persuasions, and the argument put forward by Luke Yuan that her biography would help to inspire Chinese youth in addition to publicizing her achievements, was she finally convinced.

With Yang's inspiration, I realized the importance of using an objective narrative in writing this book. As a result, I would not rely only on her own version, but would interview her colleagues, students, friends, relatives, and even competitors, and make reference to many documents and existing literature. This turned out to be a time-consuming undertaking.

I started the official interviews with Wu Chien-Shiung in New York in September 1989. They were unexpectedly rough. I had imagined a rather romantic setting, with Wu vividly recounting her life and events while relaxing in her chamber, with the tape recorder turning and my pen moving, the sun would be setting slowly. But this almost never happened in the tens of interviews in more than a year!

Wu Chien-Shiung never wrote a diary. While wholeheartedly immersed in her scientific experiments in the past years, she never considered recording this process for the world, and therefore had no memory of many events. In addition, she was down-to-earth, and of few words. The medicine lowering her blood pressure also affected her memory. The tens of hours of interviews and reminiscences were not enough to reconstruct her past.

In addition to the interviews, I have read more than ten relevant books, the personal archives of Wu Chien-Shiung in Columbia University, historical documents in the American Physical Society, thousands of pages in newspapers, magazines, and physics periodicals. Then there were the interviews with her relatives, friends, colleagues, students, scientific collaborators and competitors. I flew more than 30,000 miles all over the world to interview more than 50 individuals in China, Hong Kong, Europe, the US and Canada. The tape recording of tens of hours of these interviews was followed by telephone calls to confirm details.

Of course, the value of a book will be judged by many qualitative factors other than the above quantitative measures. The above process just illustrate how the writing of this book proceeded and developed.

To establish the credibility of the material, I have used extensive end-of-chapter notes, references, and bibliography for critical sources, comments and conversations. I also used objective narrative as much as possible, and avoided subjective opinion or novel-style description. Basically, this book conforms to the current trend in biography writing, using a narrative style closer to news reporting.

In a way, the writing of this book could be regarded as an attempt to test the idea that "news serves as a footnote of history". There must be errors and omissions due to my own limited capacity, and I sincerely welcome your corrections.

The completion of this task owes much to the contributions and extraordinary help of many individuals, including my relatives, colleagues, and many people, institutes, universities in many countries that are referred to in this book. To avoid the embarrassment of possible omission, I decided not to list their names, but to sincerely express my utmost gratitude.

Besides C. N. Yang who encouraged me to start this project at the very beginning, and gave me much direction and assistance during the course, I received the most support and assistance from Yu Ji-Zhong, CEO of the *Reading Times* where I worked. I remember seeing Yu in the spring of 1989 to propose the project of writing this biography right after getting approval from Wu Chien-Shiung. After careful consideration and probing questions, Yu immediately recognized the meaning and nature of this project, and granted me the unprecedented support of a one-year leave from the daily grinding of newspaper reporting. I was able to stay in New York City where Wu Chien-Shiung lived for long-term interviews and collection of source materials, and I was able to travel several times to other locations in the US visiting relevant individuals.

In the following years, the writing of the biography became irregular and with much delay, as I was involved with much reporting and planning back in the newsroom. Yu would always tell me not to hurry, as quality was the most important priority.

The experience of writing this biography was not unlike my newspaper work: one always felt that the material was incomplete, but had to put words down when facing a deadline. Similarly, there is a sense of lack of closure in the end, but one must do one's best, relying on whatever one can find from the recordings of her life events.

There are always conflict and choice between the pressure of time and completeness. I just hope that this book will meet the expectations of concerned and helpful individuals.

Chiang Tsai-Chien
28 May 1996

Translator's Note

To my loving wife Rosie (1944–2012)

In January 2007, Patricia Cladis was interested in the life story of Madame Wu Chien-Shiung. Surprisingly, she found no detailed biographical books written in English on Wu in the market, which remained the case until now despite Wu's tremendous accomplishments. Lui Lam introduced me to Pat, who just got hold of a copy of Chiang Tsai-Chien's book published in 1996 through her contact in Taiwan, but a book she couldn't read. By May of 2007, both Pat and Lui raised the idea of translating Chiang's book into English. After several professional translators whom we approached could not find the time, I (reluctantly) committed to give it a try.

When I did a "faithful" translation of the Chinese text, Kai Wong gave me great help correcting many mistakes in language, history, culture and physics. I also received assistance from Noemie Koller and Peter Yu, as well as Mou-Wing Heung, Shuet-Hing Lee and Odoric Wou. Koller and Yu helped me by tracking down the many names and references in physics quoted in the text in Chinese only; similarly, Heung, Lee and Wou helped me with names and references in Chinese history and nationals.

I also had many encouraging and insightful discussions with Patricia Cladis, Lui Lam, Jada Yuan and Vincent Yuan. I gratefully acknowledge the many contributions of the above individuals. I also want to thank Vincent for providing valuable photos from his and relatives' collections. Vincent and Jada made extensive changes in English and the organization of the text, which I deeply appreciate. Last but not least, I am indebted to the editors of World Scientific, especially Kostas Ikonomopoulos, Kim Tan and Roh-Suan Tung, for their copy editing of the translation draft. This effort includes tracking/referencing many Chinese persons and modern Chinese events.

However, this is Chiang Tsai-Chien's book, as it should be; he is the one who is solely responsible for the final version as printed here.

A word in the translation of Chinese names: We translated all Chinese names into pinyin (Xinhua Dictionary, Commerce Press International), unless the names (spelled differently) were well established before 1958, in which case they will be adopted. As there are many (sometimes as many as 50) Chinese characters with the same pinyin, we also created a "Name Index" with both the pinyin and the corresponding Chinese names. This should be useful for those who can read Chinese and would readily recognize the many well-known Chinese nationals quoted in the biography. In addition, there are two conventions writing the English names of Chinese, either putting the family name last, or first. We add a hyphen between the given names (usually two characters) to better identify the family names. Both conventions were used in the translation at times. The "Name Index" (sorted by family names) serves as a good reference.

This translation should make the original Chinese text accessible to English-speaking readers. We sincerely hope that the readers will find the life, struggle and career of Madame Wu inspiring, and gain appreciation of her many accomplishments.

Wong Tang-Fong, Frank
February 5, 2013

Prologue

Manhattan in New York City is a narrow granite island, measuring 20 miles from south to north, surrounded by water, with the East River to the East and Hudson River to the West. It has many skyscrapers, in which commercial and artistic activities help make New York City an ever-changing metropolis. These buildings also create a unique, charming skyline day and night.

Near the middle of Manhattan, there is an area called Harlem Heights in the 18th century. The name has been replaced by Hamilton Heights to the North and Morningside Heights to the south. Nearby Harlem is a famous older neighborhood, populated by blacks and Spanish-speaking immigrants. In general, it is an unsafe, dangerous neighborhood.

The streets in Manhattan are arranged in a basic checkerboard fashion: east-west streets are numbered sequentially with 220th street at the very North; while north–south avenues are numbered from 1st Avenue to 12th Avenue at the very West. There are some named streets and avenues mixed in between. The most famous avenue is Broadway running through Manhattan from Northwest to Southeast. The many theaters hosting stage-shows around Broadway in the busy midtown even make "Broadway Shows" a coveted label.

Going north on Broadway to about West 116th Street, one starts to notice buildings of more uniform Renaissance style with tall windows, stone pillars, red bricks and green roofs, which give rise to a different feeling. On an outside wall of a building at the corner of Broadway and West 118th Street, there is a small obscure stone plate with a sculpture of a battle scene, and a quote "Washington's Army Won the Battle Here, Harlem Heights, 9/16/1776".

This was the location where the American Army won their first victory in the Revolutionary War. According to the record, around noon at 9/16/1776, the Revolutionary Army led by General George Washington marched south in Manhattan, engaged in a furious battle in Harlem Heights with the British Army, and won.[1]

These subdued buildings in the old battlefield belong to the campus of Columbia University. The predecessor of Columbia was King's College, established by the British King George II in 1754 at a different location. This is the 5th oldest university in America and is not only located in the battlefield of the first victory of American Independence but its many early graduates were critical in advancing the independence movement.[2]

King's College moved several times, finally to the current site in Morningside Heights in 1897; it was renamed Columbia University the year before.

With its tradition and under capable leadership, Columbia became one of the few prestigious universities on the East Coast. In the 1940s, eight major universities on the East Coast, including Columbia, Harvard, Yale, Princeton and Cornell formed an Ivy League playing American Footfall matches. This league later expanded to other athletic competitions and activities.[3]

In its 238 years of history, Columbia occupied a prominent position in academia. Not only did it make achievements in many fields, it has many distinguished alumni making lasting contributions to the American society. These alumni include the founders, President Roosevelt and his father, and the philosopher J. Dewey, to name a few. Eisenhower also served as President of Columbia University for two and half years before he was elected the 34th President of the US.

Columbia University was a leader in sciences. In addition to its prominence in biology and medicine, it was a center in the physical sciences since the 1930s.

The physics department is located in the Pupin Laboratories. This building was named after M. I. Pupin, a Yugoslavian immigrant in the early 20th century, who made major contributions to science. Since its opening in 1927, there have been 17 Nobel Laureates who studied, taught or researched

◀ Campus of Columbia University.

▶ Manhattan, New York.

in this 13-floor brick building. Ten of them actually completed their physics and chemistry Nobel winning research in this building.[4]

On May 9, 1980, a gloomy and unsettling spring day of about 45 degrees Fahrenheit, there was a sense of urgency as students and professors rushed around Columbia campus near the end of an academic year.

It was not yet 10 am, people of many ages had already gathered in the Wood Auditorium, in the basement of Avery Hall (which housed the Research Institute in Architecture) near the imposing Low Memorial Library with its round-top, stone-column Renaissance style. They were there to attend the conference *Symposium in Honor of Columbia University Physicist Chien-Shiung Wu* honoring her retirement.

The Public Relations Department at Columbia issued a press release on May 5 to announce this conference, and particularly honored her as the *First Lady of Physics Research*.

The press release listed many firsts for Wu Chien-Shiung: first female President of the American Physical Society, first female recipient of the Cyrus B. Comstock Award from the US National Academy of Sciences, first female recipient of an Honorary Doctoral Degree from Princeton University, first female recipient of the Scientist of the Year award from the Industrial Research Magazine. Wu was also awarded the first Wolf Prize from Israel in 1978.

Wu Chien-Shiung was 68 years old. She had a special position among the many world-class physicists at Columbia University. She was not only a female scientist, but also a Chinese immigrant in the 1930s. She overcame the barriers in language and culture, and established a supreme position beyond gender, race and culture. She achieved great experimental results in physics in a male-dominated world.

Wu went to UC Berkeley in 1936, and retired in 1980. She published almost 200 papers in 44 years of research in physics. The achievement of a scientist is never measured by the number of publications, but more importantly by the influence of these papers in science.

The number of her publications may not be that large, but her very first paper, published in 1940 while a graduate student, was already a masterpiece with rather important influence. With an ever-maturing knowledge and experience, Wu continued to produce results with profound influence in nuclear physics.

Wu made at least three major contributions in physics: the first one was a series of precise experiments in beta decay from 1946 to the mid-1950s. This work firmly established the experimental foundation clarifying the theoretical development, and made her a world authority in beta decay research.

Then it was her rather difficult and precise experiment conducted with a group of scientists in the US Bureau of Standards near the end of 1956. This experiment was the first to confirm the hypothesis of parity violation in weak interaction as proposed by T. D. Lee and C. N. Yang. Due to such an undisputed experimental confirmation, Lee and Yang received the Nobel Prize the very following year. The third contribution was the experimental (conducted by her research group) confirmation of vector current conservation in beta decay, which incidentally was also proposed by another two Nobel winning physicists.

Many Nobel laureates and great scientists felt that any one of these achievements should have earned Wu a Nobel Prize. These include her close Chinese colleagues Yang and Lee, I. I. Rabi, E. Segrè, J. Steinberger, L. Lederman, N. Ramsey, R. Oppenheimer, and R. Wilson.[5]

Although Wu did not get a Nobel, she certainly received numerous honors. These include the Cyrus B. Comstock Award from the US National Academy of Sciences, the National Medal of Science from the US White House, and the so-called Israeli Nobel "Wolf Prize".

Wu was the Pupin Professor at Columbia, and became Professor Emeritus after retirement. In the press release, the Chairman of the Physics Department, J. Luttinger, said: "Although Dr. Wu has no intention reducing her active research at Columbia, nor her leadership role in physics, we still want to celebrate this milestone on this special day."

The attendees of the conference included Nobel Laureates I. I. Rabi, T. D. Lee, J. Rainwater, the founder of the Fermi National Accelerator Laboratory, R. Wilson, and her many collaborators, colleagues, and former and current students.

From 10 am to 3 pm, there were 10 different presentations in various topics. The speakers recounted stories of their collaborations with Wu, and their profound admiration. The former Director of Brookhaven National Laboratory, M. Goldhaber, said:" People avoid doing experiments in beta decay, simply because they know that Wu Chien-Shiung will do a better job than anybody!"[6]

There was a cocktail reception on the 9th floor in Pupin Laboratories from 4 pm to 6 pm, then a banquet in the Chinese restaurant Quan Jia Fu on Broadway. Almost 100 guests attended the dinner seated at 12 tables.

The most senior professor in the Physics Department, and Nobel Laureate Rabi, gave a long speech at dinner. He had directed the Columbia Physics Department for more than 40 years before his retirement in 1968 and was particularly close, both as a mentor and a friend, to Wu who joined Columbia in the 40s. Their residences were only a few buildings apart.

Rabi admired the achievements of Wu Chien-Shiung very much. In his speech at the retirement banquet, Rabi presented in detail her many scientific achievements, and insightful histories. Most significantly, he personally believed that Wu actually had made greater contributions to science than Madame Curie, in spite of her nickname "Chinese Madame Curie".[7]

Wu in turn had much respect for Rabi, not only for his scientific achievements, but also particularly for his courage in defending Oppenheimer in the 1950s, and for his vision of a "World Citizen". Unfortunately, this dear colleague of Wu for more than 40 years passed away in early 1988. It took away from this biography a possible valuable source of raw material and a witness. His speech in the banquet was also not recorded due to the absence of recording equipment in the restaurant.

Reflecting on her past in stages, from a young girl speaking not very fluent English, leaving war-torn China in the 1930s, arriving in the US with serious discrimination toward female scientists, Wu nevertheless became a most distinguished physicist in the 1940s, armed with her unusual persistence and unsurpassed talent. Wu Chien-Shiung further established her lasting position in physics with her world-class experiments completed in the 1950s and 1960s. This was indeed a long, fascinating journey!

The real life story and this admirable journey of Wu Chien-Shiung all started in the small village of Liuhe near Shanghai, in Jiangsu Province in China.

Notes

1. *Encyclopedia of the American Revolution*, Mark Mayo Boatner III, New York: David McKay Company Inc., 1974.

2. *A Brief History of Columbia University*, Columbia University Archive, 2/1987. The 1765 graduate Alexander Hamilton, John Jay, and Robert Livingston authored the Preface to the American Constitution. *Columbia University in New York City* booklet, Columbia Magazine.
3. *The History of the Ivy League*, William H. McCarter. The other three universities in the league are Brown University, University of Pennsylvania and Dartmouth University.
4. These Nobel Laureates are: Harold Urey, Isidor I. Rabi, Polykarp Kusch, Willis Lamb, T. D. Lee, Charles H. Townes, James Rainwater, Leon Lederman, Jack Steinberger, and Melvin Schwartz.
5. Interviews with C. N. Yang, T. D. Lee, Leon Lederman, Jack Steinberger and Glenn Seaborg, 8/1989 to 10/1990. Also based on the assessments of the late Rabi and Segrè.
6. Recording taped during the Conference honoring the retirement of Wu Chien-Shiung.
7. Interview with Wu Chien-Shiung, 6/22/1990, New York City residence.

▶

Wu Chien-Shiung in the 40s.

◀

Wu Chien-Shiung
in the 50s.

◀

Wu Chien-Shiung in approximately the 60s.

▶

Wu Chien-Shiung in the 70s.

Wu Chien-Shiung in the 70s.

Wu Chien-Shiung in the 80s.

Wu Chien-Shiung in the late 80s.

Her father Wu Zong-Yi and Wu Chien-Shiung had deep affection for each other and he had great influence on her.

In 1957, Wu Chien-Shiung, Luke C. L. Yuan, and their son Vincent in the cabin of the ocean liner to Europe.

In 1965, Wu Chien-Shiung returned to Taiwan to receive the Achievement Award, from the Chi-Tsin Culture Foundation, and attended the tea party reception of President Chiang Kai-Shek in National Defense Research Institute in Yang Ming Shan. Pictured with Soong Mei-Ling (3rd right), Chern Shiing-Shen (2nd left), Zhang Qun (left) and Yen Chia-Kan (right).

In 1964, Wu Chien-Shiung received the Cyrus B. Comstock Award from the US National Academy of Sciences. The prize was awarded by then Academy President Fredrick Seitz (left); attendees included Nobel physicist of that year Charles H. Townes (right) and US Ambassador Chiang Ting-Fu (2nd right).

◀

In the laboratory of
Columbia University.

▶

Wu Chien-Shiung and
Luke C. L. Yuan met with
Chou En-Lai in early 1973
in China.

Wu Chien–Shiung being escorted
in the White House to receive the
National Medal of Science.

▼

Wu Chien-Shiung was awarded the National Medal of Science in the White House from President Gerald Ford (1975).

During the 1977 International Nuclear Physics Symposium in Tokyo, Japan, Wu Chien-Shiung discussed with Nobel physicist Sin-Itiro Tomonaga (2nd right).

▲

In 1997, Wu Chien-Shiung, Luke C. L.
Yuan, son Vincent, and daughter-in-law
Lucy Lyon toured the Great Wall.

◀

Wu Chien-Shiung took
the office as President
of the American
Physical Society.

▶

Wu Chien-Shiung and
Luke C. L. Yuan (right)
met with the wife of the
Italian Premier Arnaldo
Forlini.

Wu Chien-Shiung, Samuel C. C. Ting (2nd right), C. N. Yang (center), and T. D. Lee (left) met with Deng Xiao-Ping in the 80s.

In 1984, Wu Chien-Shiung was awarded an Honorary Doctorate from the University of Padua in Italy. She gave a lecture in the hall where Galileo spoke.

▶

Selecting books from the shelf at home.

Wu Chien-Shiung with old friend Margaret Lewis (2nd left).

▼

◀

Son Vincent, daughter-in-law Lucy
Lyon, and granddaughter Jada.

Reunion with National Central University alumni in Taiwan. From left to right: Yu
Chuan-Tao, Li Guo-Ding, Luke C. L. Yuan, Tian Yun-Lan, Wu Chien-Shiung, Yu
Ji-Zhon, Mr. and Mrs.Yu Zhao-Zhong, Wang Zuo-Rong, Wang Cheng-Sheng.

▼

Wu Chien-Shiung and Luke C. L. Yuan, the loving couple.

▶

With granddaughter Jada.

▼

The Yuan family in front of their New York residence.

Chapter 1

Childhood in Liuhe

Liuhe, hometown of Madame Wu Chien-Shiung, is located northeast of Shanghai, about an hour away by car and along the Yangtze River.

A typical village in the "South of Yangtze" area, Liuhe has a strategic position near the mouth of the Yangtze River, overlooking Chongming Island in the middle. Liuhe is also the area's agricultural provider and a commerce center of Taicang County. It is busy and crowded with both carts and ships. Liuhe has been called "The Six Countries Pier" since the Yuan Dynasty (1279–1368), and "Junior Shanghai" indicating its prosperity. In the Yongle Year of the Ming Dynasty (1368–1644), the Grand Eunuch Zheng He departed from there for his seven naval expeditions.

In spite of China's economy opening up in recent decades, Liuhe still maintains its old world charm. It has a few little new buildings in the periphery and several new roads connecting to other cities. It still has the historical buildings in the center of town: old wooden two-storied buildings, covered with black tiles, with wooden windows, doors and ceiling beams. Stone-tablet-covered lanes crisscross the old buildings, where people open grocery stores, or set up tailor or barber shops.

There is indeed a Liu River (Liuhe means "Liu River" in Chinese) flowing by the edge of town, with peach trees on both banks, and several cement bridges across it. The water in the river is mostly stagnant and quite muddy as a result of population growth and industrial pollution.

Wu Chien-Shiung was born in Shanghai on May 13, 1912 (April 29 in the lunar calendar). Her birth brought much pleasure to her scholarly family in this village. As it was also the year of the inauguration of the Republic of China, there was a sense of renewal everywhere in the air.

2

Liuhe at present. (Photo taken by Chiang Tsai-Chien.)

Wu's father Wu Zong-Yi founded Ming De School in his hometown Liuhe. (Photo taken by Chiang Tsai-Chien.)

Wu was the second child, but the first daughter in the family. According to tradition, all children in this generation had Chien as a first character in their first name, and the second character would follow this sequence 'Ying-Shiung-Hao-Jie' (heroes and outstanding figures). Wu had an older brother Chien-Ying born in 1909, and a younger brother Chien-Hao born in 1920. When Wu Chien-Shiung was born, her grandfather Wu Yi-Feng, a junior scholar (Xiu-Cai) in the National Civil Service Examinations in the late Qing Dynasty (1644–1911), was still alive. Because her grandfather somewhat favored a male heir, Wu Chien-Shiung was never spoiled, even as the only girl in the family.

Her father Wu Zong-Yi was born in 1888, and he was 24 when Wu arrived. Zong-Yi was very progressive, knowledgeable and courageous. He was very close to Wu Chien-Shiung, loved her very much, and had a profound influence on her life.

Wu Zong-Yi studied in the Third Elementary School in Taicang County, and went on to the Nanyang Public School in Shanghai upon graduation. Nanyang Public School was founded in 1896, as proposed by Sheng Xuan-Huai, a principal in the *Self-Strengthening Faction* in the late Qing Dynasty. It had the mission to educate a new generation of Chinese to become fluent in foreign languages and technologies. It was called "Public School" as it was funded by both private businesses and by the government.

Nanyang Public School gradually expanded into various colleges, and was renamed Higher School of Commerce and Industry of the Ministry of Trade in 1905 when Sheng Xuan-Huai fell out of favor. The school had graduated three to four hundred students, and selected tens of them to go abroad for further study. Nanyang Public School was the predecessor of Jiaotong University.[1]

In such a relatively open environment, Wu Zong-Yi began his contact with the ideas of freedom and equality from the West. He read extensively, particularly books on human rights and democracy. These activities had a major effect in molding his thought process.[2]

Books and periodicals spreading new ideas were officially banned from campus. While reading these materials in secret, Wu Zong-Yi could not understand why such reasonable ideas were outlawed. This increasing resentment drove him to join the resistance movement by boycotting classes. After the boycott, Wu Zong-Yi was no longer willing to stay in

Nanyang Public School, withdrew and transferred with several like-minded students to Ai-Kuo (love your country) Academy, founded by Cai Yuan-Pei.

Cai Yuan-Pei was a widely respected revolutionary and educator. His ideal of education centered on academic freedom, tolerance and diversity. The professors in Ai-Kuo Academy were not only experts in their fields, but also invariably had revolutionary ideas. Free discussions and progressive thought flourished on campus. Various journals circulated freely and student organizations democratically elected their leaders and published their own periodicals. Cai Yuan-Pei attempted to implement a democratic system that he believed in on campus.

In 1903, the government exposed and seized the revolutionary *Jiangsu Tribune* in Shanghai. The imprisonment of two prominent revolutionaries, Zhang Bing-Lin and Zou Yong, shook China at large. Cai Yuan-Pei was implicated because of his editorial involvement with the newspaper, and was exiled to Qingtao. The academy was closed and Wu Zong-Yi discontinued his studies.

The *Jiangsu Tribune* case quieted down next year and Cai Yuan-Pei returned to Shanghai to start another newspaper. Wu Zong-Yi joined the newspaper spreading revolt against the Qing government. In 1907, he joined the Athletic Association founded by his father and others, with the mission of building a healthy body able to serve China. In 1908, Chen Ying-Shi was sent by Sun Yat-Sen from Japan to Shanghai to actively organize revolt activities. Wu Zong-Yi introduced himself to Chen as a former student of Cai Yuan-Pei, and expressed his readiness to join the revolution. He was later admitted to the Alliance Society.

Wu Zong-Yi also joined the Shanghai Merchant Corporation in 1909, to study military techniques. This group became the stronghold of the revolutionaries in 1910. It played a critical role in liberating Shanghai during the October 1911 Shanghai Revolt, which succeeded in overthrowing the Qing Dynasty and led to the founding of the Republic of China. In 1913, Yuan Shih-Kai became Acting President of the Republic. He greatly consolidated power and suppressed opponents. This act led to the Second Revolution against Yuan. Wu Zong-Yi participated in attacking the Shanghai Arsenal, and taking over the Weapon Depot. The Second Revolution unfortunately failed rapidly due to the superiority of Yuan's army, and Zong-Yi had to return home to Liuhe.

Wu Chien-Shiung was born in Shanghai during this chaotic time. The young Wu was very smart, good-looking, likable, and contemplative. She was nicknamed Wei-Wei. Wu started her education like most students of that time, with recital of poems, reading of characters, and simple arithmetic. She already showed an unusual talent in these early studies.

One day her grandfather Wu Yi-Feng called her mother: "Ping-Ping". Little Wu frowned and objected: "Grandpa, Mom is named Fu-Hua, Fan Fu-Hua, not Ping-Ping." Delighted by such an alert and naughty granddaughter, he followed: "Wei-Wei, do you know why your Mom is named Fu-Hua?" Wu Chien-Shiung pondered for a while, and said: "Yes, it was from the mission of Sun Zhong-Shan that we should drive away the foreign occupiers and *rebuild China* (Fu-Hua). Dad always said that Sun is a good man, so he changed Mom's name to remind us of the mission." Up to that point, Wu Yi-Feng had felt disappointed in having a female grandchild; he finally gave up the prejudice and increasingly treasured his only granddaughter.[3]

Her father was without doubt the biggest influence in the development of the young Wu Chien-Shiung. In her memory, this period of growing-up was a happy one. The three siblings read a lot, always encouraged by their father. Wu said that her father was always ahead of his time, inquisitive and hungry for learning.[4]

Wu Chien-Shiung said her father was indeed very progressive. Even before the founding of the Republic of China, he recognized that the time of Empress Dowager (Cixi) and her heirs had passed, and that there would be major changes in the East. He wished his children to be suitably prepared to live in a modern world. This would require a profound understanding of the value of Chinese culture, enabling them to have a meaningful and rich life.

Not only was Wu Zong-Yi progressive, he had broad interests and was quite accomplished in radio, hunting, playing the accordion, singing, and reciting classical poems. Though he was not demanding of his children, he gave Wu an extra push because he recognized her quiet curiosity and unusual talent. He frequently read aloud the scientific news articles in the Shanghai newspaper, "Shen", to her (before she could read). The primitive quartz radio that he put together really fascinated her and she was intrigued by the ability of the radio to receive news from far away.

The Wus lived in an old, two-story house to the right of the plaza in front of the Tian-Fei (Goddess) Temple in central Liuhe. Wu would play with other children in the plaza, but preferred to stay home listening to the radio. She was quieter and more thoughtful than other children her age.[5]

She received her elementary school education at the Ming De School founded by her father. She was proud of and inspired by her father's process in founding the school, his courage and his introducing of modern concepts to the village.

When Wu was very young, Liuhe was not peaceful. There was a group of bandits who preyed on the village, robbing and kidnapping. The locals could not do much. Even the police department was looking the other way.

Because Wu Zong-Yi had received military training in the Shanghai Merchant Corps and had participated in attacking the Shanghai Arsenal during the Second Revolution, he founded a local militia upon his return to Liuhe, and developed a plan to eliminate the bandits by executing their leader.

The leader of the bandits was Wang Ying-Biao. Wu Zong-Yi monitored their routine and selected the right time to invade their headquarters. He led the attack and personally killed Wang. Without a leader, the gang dissolved and disappeared from the area.

With the return of peace, Wu Zong-Yi turned his attention to education. There was a Temple of the Fire God that had fallen out of active service because of the turmoil caused by the bandits. Its courtyard was used for training the local militia. He gained the support of the local men, removed the clay statue of the Fire God, and renovated the temple into a school. The arched entrance gate proudly proclaimed "Ming De Women's Vocational School" — a name taken from the proverb: "The journey of education is in understanding morals (Ming De's direct translation in English means "understanding morals")". Zong-Yi served as the school principal.

The school had only a few children from close relatives at the beginning. Wu Zong-Yi would go door to door with Chien-Shiung to recruit students, rich and poor. There was no charge for attending the school. It would teach sewing, embroidery, gardening, and similar practical trades. Zong-Yi wanted to eliminate ignorance through education, and prejudice against women — embedded in the old saying "Moral is an uneducated woman." Ming De School even admitted girls obliged to care for their brothers — with the brothers attending the school too.

Not only was Wu Chien-Shiung hungry for learning, it is clear from interviews that her father's personality had great impact on her development and education. Her persistence in research, her principles and her integrity were critical in her emergence and eventual prominence in science.

Wu remembered that her father was actually quite successful in modernizing the village, despite the villagers' resistance to his progressive thinking.

For example, when Wu Zong-Yi removed the clay statue of the Fire God during the school's renovation, the locals were not happy. Zong-Yi arranged an elaborate parade escorting the statue to another temple. Such a gesture earned much support from the locals, despite the fact that he worked for foreign companies in Shanghai, was fluent in English, and did not really share the local belief in the Fire God.[6]

The quartz radio was another of his hobbies. Wu Zong-Yi built many radios and gave them to the local homes, and to all teahouses in Liuhe. Because the teahouses were the gathering place where locals spent their leisure time, this provided them an opportunity to reach the outside world. He rented movies in Shanghai in the summer, which also helped educating the locals.

He used modern materials and the Chinese phonetic system in the Ming De School, and continued to send Wu's mother and aunts to recruit girls in the village. Over 50 girls (including Wu) from the small village Liuhe further pursued their education in Shanghai and Suzhou. Not a small feat![7]

Wu Zong-Yi had a younger brother called Zhuo-Zhi, also an energetic and progressive young man. They both joined the Second Revolution. Wu Chien-Shiung liked to recall that they attacked the Shanghai Arsenal day and night, except for a short dinner break. Zhuo-Zhi went to France in 1919 to take part in the "Work-and-Study Program".

They later founded the Hutai Transport Company based in Liuhe, managing the transportation of goods between Taicang and Shanghai (short name of Shanghai is Hu). They trained their own drivers to maintain safety and quality of service. Hutai Transport Company later assisted in the transportation of personnel and supply of the 19th Army in 1931, after the Japanese army landed in Shanghai in January 1928. It also supported the government army and achieved early victory in the August 13 Song-Hu Battle after the outbreak of the Sino-Japanese War in 1937.

Wu admired her father and uncle for their energy, their knowledge, and their vision for the cultivation of a new China. She felt the subtle restrictions as a female, but was determined to accomplish like her father and her uncle; like — a man.

Wu received her elementary, formal education in Ming De School. She studied relatively modern subjects such as arithmetic and the Chinese phonetic system, in addition to the classics such as *The Analects of Confucius* (Lun-Yu) and *Collection of Ancient Literature* (Gu-Wen-Guan-Gi). The phonetic system was designed to help standardize the pronunciation of Mandarin when people spoke different dialects in different provinces. Wu spoke fluent Mandarin, but with a rather heavy Shanghai accent.

In her memory, it was her father's daily practice that influenced her the most. Wu and her brothers always read and discussed various topics together with their father. She remembered that one day her father asked them to locate a stone tablet in the courtyard of the Tian-Fei Temple. He explained that the tablet recorded the journal of the Grand Eunuch Zheng He in his far-reaching explorations.

In her father's account of the explorations, Wu was most impressed by the huge size (in both number and tonnage) of the fleets, and the diplomacy that Zheng He used in dealing with many cultures in different countries. For instance, some sailors were converted to Islam while visiting the Middle East and wanted to stay; Zheng He would grant their wishes and agree to pick them up on the next trip.

Wu often retold these stories of clever ways in managing different cultures, and was profoundly inspired by them.

At this time, the Tian-Fei Temple in Liuhe had been converted to the Zheng He Memorial. The wooden building housed maps, records, and ship models used in the explorations, as well as portraits of Zheng He. The temple was located behind Wu's family house, which sadly no longer exists.

Wu Zong-Yi worked on introducing new knowledge to Wu. He frequently read the Junior Encyclopedia (Shanghai Commercial Press) and described stories of scientists to Wu. Wu Chien-Shiung was always fascinated by the research and discovery activities of nature.

◀

With uncle Wu Zhuo-Zhi as a youngster.

▼

With father Wu Zong-Yi, mother Fan Fu-Hua (left), and brother Wu Chien-Ying (2nd left).

In terms of personality and principles, her father used more of the Chinese tradition in teaching Wu. She recalled that her father never lectured her directly, but used examples, when appropriate, in classical Chinese essays to illustrate given situations. This helped her to have a better appreciation of the essays. He always said that there were gems in these essays.

While her father was eager to adopt modern concepts, he also appreciated the value and beliefs of Chinese culture. Similarly, for the rest of her life Wu would always uphold the merit of Chinese culture and tradition, even when she was at the frontier of modern science. She remembered her father's teaching — the persistence and can-do attitude in the face of difficulty.

Wu had a happy family life as a youngster. She once said that she just could not understand the conflicts between parents in other families. In her own family, her parents were always working together with her mother helping out in the school.[8]

It was chaotic and full of unrest outside of Liuhe during the first eleven years of the Republic of China. Wu was fortunate to stay in Liuhe, and enjoyed a happy chidhood. It was a good, protective environment, which prepared her to depart for (Soochow) Girl's High School, fifty miles away from home at the age of eleven.

Notes

1. *History of Jiaotong University*, Chapter 1: The Era of Nanyang Public School (pages 1–11), Shanghai Education Publishing Co., 1986.
2. Highlights of the life of Wu Zong-Yi in this chapter were mainly taken from *Biography of Wu Zong-Yi*, Yang Gong-Huai.
3. This story was taken from "*Wu Chien-Shiung — A modern day Madame Curie*", Pang Rui-Gen, Hunan Literature and Art Publishing Co., August 1987. Wu confirmed the facts.
4. *Current Biography Yearbook*, Ed. C. Moritz, H. W. Wilson Company, New York, 1959.
5. Interviews with Wu Chien-Shiung's cousins Hong Wan-Zhen and Zhang Xie-He, Taipei, May 1989, and Geneva, Switzerland, August 6, 1989.
6. Interview with Wu Chien-Shiung, September 13, 1989, New York City residence.
7. Same as 6.
8. Interview with Wu Chien-Shiung, June 22, 1990, New York City residence.

Chapter 2

A Young Wu Chien-Shiung Became the Favorite Student of Hu Shih

In 1923, at age 11, Wu Chien-Shiung left Liuhe to continue her education in Suzhou, 50 miles away. Suzhou is a beautiful, almost heaven-like city. As the saying goes: "Suzhou and Hangzhou are heaven on earth."

Wu took the matriculation examination for the Soochow Girl's High School (Second Women's Normal School) in Suzhou. That year, the school admitted two classes of teachers-in-training (normal school) and two classes of regular high school students, a total of 200. Wu ranked ninth among some 10,000 applicants.

Wu applied for the teachers-in-training class, which did not charge for tuition or room and board, and guaranteed a teaching job after graduation. In fact, Wu's family was fairly well-off and could afford the expenses of the regular high school. But the teachers-in-training program was more prestigious then, and Wu wanted the challenge of a difficult admissions process and a competitive program.[1]

The Soochow Girl's High School in Suzhou was a rather famous high school. The principal Yang Hui-Yu was a forward-looking educator, with many contemporary ideas. Her experimental education programs earned the school a great reputation in the education community. Many educators, both domestic and foreign, visited the school for ideas. Wu was aware of this unique feature when she applied.

In their experimental education programs, Soochow Girl's High School employed many excellent teachers, and used contemporary materials. They frequently invited famous scholars, domestic and foreign, to present distinguished lectures.

Wu remembered that those lecturers included the great American philosopher and educator J. Dewey; his first student W. H. Kilpatrick; and

P. Monroe. The Dewey School had a lasting impact on American educational theory, emphasizing a concept of education grounded in real life experiences. This philosophy greatly expanded Wu's view on education and discovery. She also recalled the lecturers' visit to her Home Education class, when the class cooked dinner for the visitors.[2]

Dewey was a long-time professor in the Teachers' College at Columbia University, and lived in China for two years, beginning in 1919. When Wu listened to their lectures, she did not imagine that she would join Columbia University twenty years later, and retire from there after a tenure of 36 years.

One other lecture, given by the young Peking University professor Hu Shi, had a profound impact on Wu. Before Hu Shih's visit and lecture, Wu had already been to the library to study his many articles appearing in magazines such as *New Youth* and *Nu-Li [Making Great Efforts] Weekly*. But what she admired most was his modern ideology. Hu Shih was American-educated, but had returned home to attempt to reform old China.

Wu Chien-Shiung had been at Soochow Girl's High School for some time when Hu Shih visited. Principal Yang Hui-Yu knew that Wu was a good writer, and also knew that she greatly admired Hu. She asked, "Chien-Shiung, since you have always been fond of Mr. Hu's ideology, could you record the lecture this time?"

The title of Hu's lecture was "Modern Women". It discussed the necessary thought process that women should follow to free themselves from the old traditions. Wu particularly remembered Hu's example in the lecture of a dirt-poor Chinese old lady. The lady had to survive, and never returned money or valuables if she happened to come across them scavenging in the garbage. Hu argued that moral standards are not absolute, but linked closely to living conditions. Wu found this more objective assessment of human nature, instead of the old school of absolute moral standards, inspiring.[3]

So inspiring, in fact, that Wu went to attend another Hu lecture at Dongwu University the following day. This time Hu Shih discussed the topic of women in modern times and social reform. The teenager Wu found these new ideologies stimulating and exciting.[4]

Wu was young and petite, but her tiny frame belied a brilliance that soon earned the attention of fellow students. She became a favorite of her teachers. Another student, Wu Zi-Wo, a couple of years older than Wu, wrote an article about student life then, saying that the seniors all wanted to have fun with the "little radish". "Radish" was a nickname given to all freshmen, who, irrespective of their age and height, were required to wear skirts and pull their hair together in a bun. They all looked like old ladies, or "little radishes" in appearance. The fame of Wu caught the attention of these seniors; and she was the youngest, cutest "little radish" of all freshmen.[5]

Soochow Girl's High School was a boarding school, with six students to a room. During Wu's junior year she met a particularly naughty senior called Shi Ren-Fan, who later wrote under the pen name Yi Fan and became the maternal grandmother of the internationally-known actor Joan Chen. Shi was very resourceful at smuggling in junior students and Wu Chien-Shiung was one of the juniors who frequently visited upper-classmen in their classrooms, becoming good friends with both Wu Zi-Wo and Shi Ren-Fan.

Word of Wu's academic talent spread very quickly on campus. In a dinner conversation one day, Wu found out that the curriculum in the regular high school had more science and English offerings than hers in the teachers-in-training program. Wu, who had an unusual talent for the sciences, was so eager to learn that she immediately worked out a scheme: She would borrow science textbooks from friends in the academic school program, and teach herself mathematics, physics and chemistry at night.

Yen Mei-He, a student one year older than Wu, remembered that other students would often remark on Wu's successful studies and progress. They claimed that Wu sometimes would not go to sleep if she was struggling to solve a mathematics problem.[6]

Yen and Wu had entered Soochow Girl's High School the same year, and both went on to National Central University, with Yen majoring in chemistry. Yen did postgraduate studies in Germany, returned to China, and taught in Beijing Normal University for 30 years until her retirement. She was visiting her son, who was studying in the U.S. in 1990, and fondly reminisced about Wu when we met in New York City. Yen, who sadly passed away in China two years later, was 80 at the time, but remarkably clear in memory and energetic in speech.

Yen Mei-He recalled that Wu was young, petite, quiet, and very bright. There were three other girls in the freshmen class, all petite and bright. Because these four sat together, forming a perfect square, they were referred to as the "Four Bean Curds".

The other three "Bean Curds" were Hua Qiao, Yao Zi-Zhen and Ge Bang-Yong. They were dear friends of Wu, and had interest and talent in literature. Hua went on to teach high school math in Shanghai, Yao became a famous writer with the pen name Luo Hong. There was another friend, Xin Pin-Lian, who became a university professor in Shanghai teaching Chinese Classics.

Wu herself was a good writer. She preferred the contemporary instead of the classical style. Wu once wrote a report of a track-and-field meet, which her father had attended when he visited her, that was so good that her teacher in Chinese Classics praised it highly and used it as a writing sample for the class. Wu was obviously very pleased. She laughed: "I was very proud of it. Hey, I did a good job!"[7]

She also remembered another article that was highly praised in the class. The teacher commented: "Very powerful — the pen writes like a forceful trunk; very farsighted — eyes so much above head." Wu was so pleased she purposely left the article and comments for her father to see while she was home on vacation. Her father did not say a thing.

Wu was curious and asked her father: "Have you seen my article? What do you think?" Her father said: "It was nicely written, but there were holes in your argument. It is very important to have substance in any article!" This remark had a lasting impact on Wu.[8]

She had talents in Chinese literature and history, and had many like-minded friends at Soochow Girl's High School; but she gradually migrated to a different path of science and discovery. From her reading and study, she was greatly excited by the new discoveries in the world outside of China and in the revolutionary knowledge about the universe resulting from developments in European science. Most inspiring to Wu as a teenager was the biography of Madame Marie Curie, who became world-famous for her discovery of radioactive elements in France. This understanding and interest eventually molded her life-long pursuit of physics.

Wu differed from her classmates not only in her path in scientific research, but also in her personality — a strong determination to achieve, and a profound insight of the world. She loved reading the biography of the French general Napoleon Bonaparte. She said that Napoleon was not only a great general, but also had genuine feeling and concern toward his army as shown in his many speeches to his subordinates. She felt this was the true secret to his success.[9]

Because Wu studied hard at school, she would skip many school trips, although once she did go on a trip to a temple in Suzhou. There she received a fortune stick that said, "Pine trees that are nurtured in the cold seasons will have great success in the future." She kept it in mind for all ensuing years.[10]

Wu studied in Soochow Girl's High School for six years (1923–1929). By the end of those six years of learning and life experience, she had transformed from a budding child to a knowledgeable, levelheaded teenager. From the descriptions and photos at that time, Wu as a very pretty teen who nonetheless projected an unmistakable ambition and confidence even with her modest dresses and proper manner. For a seventeen-year-old, the future is full of opportunities and challenges.

Wu graduated with the highest grades in her class in 1929, and was recommended (with no need for entrance examination) to attend the National Central University in Nanjing. Because she was in the normal school program, she was supposed to teach for one year upon graduation before she could attend university. But in her case, the mandatory teaching regulation was not quite enforced. Wu instead attended National China College in Shanghai, and became the favorite student of Hu Shih — the prominent philosopher whose lectures about women's place in the modern world had so influenced Wu in her early school years.

National China College was the first private university in China. In 1906, when the Chinese students in Japan could no longer tolerate the discrimination practiced there, they collectively withdrew and founded their own university in Shanghai. Hu Shih studied there. He resolved the conflict and the student boycott in the university in 1928, and then took a joint appointment as the President of the university, in addition to his post at Peking University. As the President, he implemented many enhancements, recruited many first rate professors, and taught a course on the History of Chinese Culture.

After her graduation from Soochow Girl's High School, Wu returned home. Her mother was thrilled to spend some time with her daughter after all these years. Her father was happy, too, but told her that the excellent professor, Hu Shih, was going to lecture in the summer at the National China College, and asked if Wu was interested in auditing the course. Wu was excited, but her mother did not like the idea.

▶

With parents.

▼

With father and family members.

Her father then suggested: "The whole family should go and have a picnic in Wusong, and then send Chien-Shiung to National China College." She admired the way her father handled such a delicate problem, and said that the two most influential men in her life were her father and her dear teacher Hu Shih.[11]

Wu was always hungry for learning, and felt that there were not enough courses (in mathematics, physics and even liberal arts) offered at Soochow Girl's High School. At National China College, she took two courses in mathematics and three courses in the humanities — a history course by the historian Yang Hong-Lie, a sociology course by the social scholar Ma Jun-Wu, and the most beneficial course "History of Ideologies in the 300-Year Qing Dynasty" by Hu Shih.

Hu Shih was a famous professor in China. He attained a rock star status with both his charisma and his scholarship. His lecture at National China College was given once a week, and lasted for two hours non-stop. Because so many students attended, the lectures were delivered in the auditorium.

Wu was attracted to Hu Shih's charisma, and most impressed by the animation of his delivery and his refreshing concepts. She still had pleasant memories when she recounted these lectures many years later.

Hu Shih took notice of Wu probably after the first examination. In that three-hour examination, Wu sat at the front row center, right in front of Hu. She finished in two hours, and overheard Hu's comment: "Well, we had this girl sitting up straight and writing non-stop. She finished in less than two hours."

Hu Shih quickly graded the paper and reported to the office. Both Yang Hong-Lie and Ma Jun-Wu happened to be there. Hu said he had never met a student with such a profound understanding of the ideologies in the Qing Dynasty, and gave her a perfect score. Yang and Ma also said a girl in their courses always got perfect scores. They wrote down the name independently, and it was the same: Wu Chien-Shiung.[12] This special bond had a lasting impact on Wu. Hu Shih also said in public that it was the most enjoyable, and the proudest achievement of his life.[13]

In addition to the regular courses, Wu also took a writing course with the famous author Huang Bai-Wei. Wu remembered being struck by the plight of a poor boat family on her way to Wusong. She wrote an assay on their tough life, and sent it to Huang.

Huang read the essay and was deeply moved. She wrote a letter full of encouragement, and had it specially delivered to Wu in her dormitory in Wusong. Wu was not home, but a friend of her roommate, Zhang Zhao-He, forwarded the letter to her. Zhang studied at National China College for four years, and later married her professor — a well-known scholar in literature, Shen Cong-Wen, who had been recruited by Hu Shih.[14]

Although Wu ended up doing scientific research, her writing was outstanding thanks to her early training. Her Chinese calligraphy was very good too, complimented by her friends as, "Energetic and elegant, without a touch of feminism."[15]

Before matriculating to National Central University Wu spent the summer preparing for her studies with her usual full force. She felt that her background and training in Soochow Girl's High School were insufficient to prepare her for majoring in science.

Her father encouraged her to plunge ahead, and bought her three books for her self-study that summer: Trigonometry, algebra, and geometry. This experience was the beginning of her habit of self-study, and gave her sufficient confidence to major in mathematics in the fall of 1930.[16]

In this period of growth as a teenager, Wu received much nurturing from many people. The most profound impact unquestionably came from her father and Hu Shih. With respect to her father, Wu remembered not only his contemporary thinking, but also his genuine respect for her. He always started any request with, "Please." He was progressive even though he had never been abroad.

After Wu left China for America, father and daughter sadly never saw each other again. In all the subsequent years, Wu rarely met people comparable to her own father.[17]

Hu Shih's influence was long lasting. In addition to the year he taught her at National China College, Hu and Wu met many times afterwards in China and in the U.S. Hu valued her, and had very high expectations of her. He once noticed in his travel a Collection of Letters of the great British physicist E. Rutherford, and figured that Wu as a physics student might be interested. He bought the book and mailed it to her. She kept this kind of mutual admiration in her mind, saying: "This is a proper student-teacher relationship!"[18]

In 1936, Hu Shih came to the U.S. to attend and lecture in Harvard University's tercentennial celebration. He visited UC Berkeley on his way home at the end of October, and had a long dinner discussion with then-graduate student Wu and others. While waiting for his ship returning to China the following day, he wrote a long letter filled with encouragement and expectation:

Ms Chien-Shiung,

I was so happy and pleasantly surprised to meet you in the Ma residence last night. In this meeting abroad, I was happy with your determination. Although our discussions were rather random, they actually had value coming from experience. Successful scholarship demands talent in addition to persistence. The example of the race between a hare and a turtle encourages persistence (turtle) and warns talent (hare). Talent and persistence are both necessary to form a winning combination. Persistence alone will do all right, but never lead to great success.

You are a brilliant girl. Do treasure yourself and you will succeed without bound. Actually this is not what I want to say. I do want to urge you to take advantage of your stay here, to understand the culture, to read books on science and other subjects, and become a Renaissance man. I do not mean to "lure" you back to the humanities; I just want you to be a broad scholar. When I met Dr. Robert M. Millikan in Pasadena a couple of days ago, he showed me around in several research laboratories. I was impressed when he demonstrated his detailed understanding even of the Genetics Laboratories. All first-rate scientists have broad interests and understanding, which also contribute to the success in their own field.

Several senior leaders in China such as Ting Zai-Jun and Yong Weng-Ni are all very broad; their leadership roles are not limited to geology. On the other hand, the young scientists are often just experts in their own fields, will never be able to create a new arena.

This is my expectation. Hope you do not feel that I am too nosy? Wishing you the best.

Hu Shih, October 30, 1936[19]

After some ten days, Hu realized that he mistakenly wrote the middle initial of Millikan as "M", instead of the correct one, "A", in the letter. He wrote a short note and asked Wu to correct it. This attitude of absolute accuracy also inspired Wu.[20]

On May 1, 1959, Wu had been in the U.S. for 23 years, and had many world-class accomplishments in science. She wrote a letter to Hu, in which a paragraph read:

> When clearing up the old correspondence several weeks ago, I came across the letters you sent me when I was a student on the West Coast. There was this one you wrote me soon after my arrival. I appreciated very much the encouragement, expectation, and advice in the letter. I made a copy for your reminiscence. Do you remember it? In my life my father and you influenced me the most. My father sadly passed away on January 3 of this year in Shanghai, and my brother Chien-Ying also passed away last June. Knowing that we can no longer see each other is so heartbreaking that I cannot hold back my tears anymore ..."[21]

Notes

1. Interview with Wu Chien-Shiung, June 21, 1990, New York residence. *Dr. Wu Chien-Shiung and Scientific Research*, Wu Zi-Wo, in *Tribute to Dr. Wu Chien-Shiung on the occasion of receiving the Special Contribution Award from Chia Hsin Cultural Foundation*. Page 17, July 1965, Taipei.

2. Interview with Wu Chien-Shiung, June 22, 1990, New York residence. Also based on a lecture she delivered when accepting the *Lifetime Achievement Award* from Radcliff College, Harvard University in 1983.

3. Interview with Wu Chien-Shiung, June 22, 1990, New York residence.

4. Interview with Wu Chien-Shiung, December 5, 1989, New York residence.

5. *Dr. Wu Chien-Shiung and Scientific Research*, Wu Zi-Wo, in *Tribute to Dr. Wu Chien-Shiung on the occasion of receiving the Special Contribution Award from Chia Hsin Cultural Foundation*. Page 17, July 1965, Taipei.

6. Interview with Yen Mei-He, June 13, 1990, New York City uptown.

7. Interview with Wu Chien-Shiung, June 22, 1990, New York residence.

8. Interview with Wu Chien-Shiung, April 22, 1993, Yuan-Shan Hotel, Taipei.

9. Interview with Wu Chien-Shiung, June 22, 1990, New York residence.

10. Interview with Wu Chien-Shiung, April 22, 1993, Taipei.

11. Interview with Wu Chien-Shiung, December 5, 1989, New York residence.

12. Interview with Wu Chien-Shiung, December 5, 1989, New York residence.

13. Speech of Hu Shih delivered at the Academia Sinica Banquet, February 24, 1962. Hu died of heart attack soon after delivering the speech. *Draft of a Chronicle History of Mr. Hu Shih*, ed., Hu Song-Ping.
14. Interview with Wu Chien-Shiung, August 28, 1990, New York residence.
15. *Wu Chien-Shiung — A Chinese Madame Curie*, Sun Duo-Ci, in *Tribute to Dr. Wu Chien-Shiung on the occasion of receiving the Special Contribution Award from Chia Hsin Cultural Foundation*, page 20, July 1965, Taipei.
16. Interview with Wu Chien-Shiung, June 22, 1990, New York residence.
17. Interview with Wu Chien-Shiung, June 22, 1990, New York residence.
18. Interview with Wu Chien-Shiung, August 28, 1990, New York residence.
19. *Draft of a Chronicle History of Mr. Hu Shih*, ed. Hu Song-Ping, Volume 4. In this letter, *Ting Zai-Jun* and *Yong Weng-Ni* were aliases of Ting Wen-Jiang and Yong Wen-Hao respectively. The *Geretic* Laboratories may be a mis-spelling of Genetic.
20. *Draft of a Chronicle History of Mr. Hu Shih*, ed. Hu Song-Ping, Volume 4.
21. The original of this letter was returned to the wife of Hu Shih, Ms. Jiang Dong-Xiu in July 1965 when Wu returned to Taiwan to accept the Special Contribution Award from Chia Hsin Cultural Foundation. It is now preserved in the Hu Shih Memorial in Nangang.

Chapter 3

Choice of Future

When Wu Chien-Shiung entered National Central University in 1930, it was a chaotic time in modern China. While the country was somewhat stable domestically due to the success of the Northern Expedition, the relationship with Japan had greatly worsened to a breaking point.

Several educators founded National Central University in the Dashiqiao (Stone Bridge) district in Nanjing. The University was called Nanjing Senior Normal in 1915, and the first President was Jiang Qian, a well-known scholar in classics.

While Jiang was President, the provost, a great educator called Guo Bing-Wen, ran the day-to-day affairs. Guo was the first Chinese to earn a Ph.D. from the Teacher's College at Columbia University, and returned to China in 1914.

Guo understood the importance of professorship to a university. He developed many contacts when serving as the President of the *Chinese Students Union in US*, and he made a special trip to Europe and the U.S. for recruiting. Guo was able to put together a team of many first-rate professors very quickly. The former President of Yenching University, and later US Ambassador to China, Leighton Stuart, said in his memoir: "All the fifty or so professors recruited by Guo were experts in their fields." Actually, the most famous professor in China, Hu Shih, would have gone to National Central University if not previously committed to Cai Yuan-Pei of Peking University.

Guo became President in 1919 and started the next phase of planning. He expanded the school to a full-blown university and named it Southeast University in 1921.

For political reasons, Guo was forced by the Northern Warlords Government to resign in 1925. By that time, under his leadership, Southeast

University had established an agricultural school in Ting-Jia-Qiao, an observatory in Beiji (North Pole) Mansion, and a fishery station in a corner of Xuanwu Lake. It was a far cry from its humble beginning as Nanjing Senior Normal with locations limited to the Si-Pai-Lou area.

National Central University experienced a period of chaos — student boycotts, frequent changes of presidency — before Wu's matriculation. Due to the able professors whom Guo had recruited, National Central did not stall, but developed into a prestigious university with a great academic environment.[1]

National Central University was a boarding school, with its women's dormitory in Shi-Po-Po Lane, outside of the back gate and at the foot of the scenic hill holding Beiji (North Pole) Mansion. The women's dormitory had four buildings that used to belong to the missionary. The rooms had various sizes, and housed one, three or six students. Wu lived in the South Building. She first shared a room with others, but eventually moved to a single room so that she could concentrate on her studies.[2]

The auditorium, library, and science buildings were all of Renaissance style with stone pillars in the front. An old (Six Dynasties, 420–581 A.D.) pine tree stood in the middle of campus. Wu started her journey in science in this European style campus, but while at the frontier in Chinese scholarship.

Wu majored in mathematics as a freshman, but transferred to physics the following year. There were many good professors there in the physics department, such as the Chairman Fang Guang-Qi who was doing research on optics. Fang, an imposing man, was enthusiastic about nurturing young physicists, and was exceptionally supportive and encouraging of this brilliant female student Wu. There was Shi Shi-Yuan, who worked for Madame Marie Curie in Europe, Zhang Yu-Zhe who became a famous astronomer and the Director of the Zi-Jin-Shan (Purple Hill) Observatory in Nanjing, and Ni Shang-Da who taught electromagnetism. All were first-rate scientists.

Anyone who followed the development of modern physics could understand why it had attracted the interest of a talented, budding scholar like Wu. When she began her university studies in the 30s, modern physics had just undergone a revolutionary development in Europe.

Western science has a tradition of continually probing the most fundamental structure of matter. Near the end of the 19th century, the important discoveries of X-ray, the electron, the Zeeman effect, and radioactivity shed light in the further understanding of atomic structure. The pioneering experiments of Wilhelm Röntgen, Henri Becquerel, the Curies, J. J. Thomson, Ernest Rutherford and others constituted a solid foundation for the coming conceptual revolution in the 20th century in physics.

In 1900, the German theoretical physicist Max Planck proposed a solution to the "black body radiation" problem. With further investigation from Albert Einstein, the concept of a "quantum" emerged. In addition, Einstein's Theory of Relativity, and the emergence of quantum mechanics in the 1920s, all helped to shake up the classical physics hitherto regarded as fairly completely developed in the 19th century. Thus, physics began another phase, and reached new heights.[3]

Wu Chien-Shiung at 18 might not have fully understood these fascinating and astonishing developments, but she definitely knew of Madame Curie's research from her earlier readings. Her classmates at National Central University said Madame Curie was Wu's role model. Wu would show her respect whenever Curie was mentioned. Wu never imagined that she would be honored as the "Chinese Madame Curie" in less than 20 years, based on her distinguished experiments in nuclear physics. Her teacher even believed that her scientific contributions were greater than Madame Curie's.[4]

Wu was very serious. Her classmates all remembered that she would struggle alone and not go to sleep until she solved a problem. Her good friend, chemistry major Cheng Chong-Dao wrote a memoir of Wu. They both lived in the South Building, with Cheng on the second floor in front, overlooking Wu's room on the ground floor in the back. Cheng wrote that Wu always stayed in her small room — not quite 100 square feet, just enough space for a table, a chair, and a bed — studying and, sometimes even reading by a flickering candle light after the dormitory electricity was turned off.

Her friends said she was brilliant, but never arrogant or showy as a young genius. They all agreed on an image of a contemplative, warm, and serious student. In her student days, Wu never bragged about the fashionable scientific theories, but diligently worked on problems at hand. This became her research style in the future.

▲

▶

Graduation from National Central
University.

Wu did not participate in many activities outside of class. Her dear uncle Wu Zhuo-Zhi would sometimes drive her and friends around town, or take them to dinners on holidays; but Wu generally would rather stay behind at school. She was by no means a loner, and made new friends, in addition to her old buddies, Wu Zi-Wo and Shi Ren-Fan, from her Soochow Girl's High School days. She was elegant, lively, gentle, petite, and lovely, naturally well liked by, and close to many female students.[5]

The contemporary Chinese painter, the late Sun Duo-Ci, who was also a student at National Central University, wrote that Wu was a discussion topic among students — she was pretty, proper, and talented. One day, Sun mentioned to another art student Liu that Wu was her favorite student, even though she had only known Wu by reputation. Liu (who came from the same village as Wu) jumped up in joy, saying that Wu had said the same of Sun.

Liu dragged Sun to the South Building one evening and "officially" introduced them. The formality made it rather uneasy whenever they saw each other on campus. But their mutual admiration led to a lasting friendship and bonding in later years.[6]

Wu had her own character with friends. Another art student Zhang Xi-Ying said Wu would ignore, but never offend, people whom she did not care about. Wu was good-natured and never showed temper. She just kept quiet when she was really unhappy.[7]

When her good friends were in trouble and in need of advice, Wu could be brutally honest. For example, when her dear friend Sun Duo-Ci was romantically involved with her professor, the great painter Hsu Pei-Hung, Wu advised Sun bluntly that she should not be too submissive. Sun would just make the problem messy by not facing the relationship.[8]

The straightforward advice she gave Sun demonstrated Wu's no-nonsense personality. In the memoir that Sun wrote some 30 years ago, she remembered running into Wu when she went to the university office to pick up their respective scholarships. Wu said: "It is so nice to see you here!" with a subtle hint of encouragement. Sun also wrote that Wu acted like a big sister, as she was one year senior. She never criticized others. Wu was always her idol, and a dear friend.

Although Wu majored in physics, and her talent in mathematics and physics were highly praised, she never felt that science was superior. She had many friends in fine arts, and she was interested in fine arts and humanities. In discussions with them, she also expressed her unique opinions.[9]

At that time, locals in the Jiangsu area liked to make fun of people from nearby Yangzhou, thinking that their frequent expression "like this, like that" was too folksy. Wu once jokingly asked Yen Mei-He, if an elegant lady like Lin Dai-Yu in *The Dream of the Red Chambers* would also use the expression "like this, like that". They both laughed heartily.

Yen later found out that Lin Dai-Yu was not born in Yangzhou, but her father was briefly there as an officer. This showed that the otherwise serious Wu also had her humorous side.[10]

Wu had several other good friends at National Central University. One was Zhu Ru-Hua, who majored in chemistry, and later became successful in the US. Another dear friend was Dong Ruo-Fen who came from the same village as Wu. They later took the same ocean liner to the US.

Another special girl friend was Cao Cheng-Ying. Cao was the step-cousin of Hu Shih. She was ten years younger than Hu, also from Jixi, and a very talented girl.

When Hu Shih was recovering from an illness in West Lake, Hongzhou in 1923, Cao was studying in Hongzhou Women's Normal University. The proximity, and the similarity in interests and talents drove them to develop a romantic relationship. It was rumored that the famous poem *Feeling:* "*The hard shell is unbreakable, but will yield when the seed inside is ready to grow. Similarly, a wall a hundred feet tall, or a tradition a thousand years old can never lock up a youthful love!*" was dedicated to Cao.

This relationship did not end well. Cao divorced her husband from an arranged marriage, but Hu could not get a divorce from his screaming wife Jiang Dong-Xiu. Cao was heartbroken, and devoted her energy to study. She attended the Agricultural School in National Central University after graduating from Hongzhou Women's Normal.[11] Cao was a Teaching Assistant in the Agricultural School when she first met Wu.

Cao was very nice to Wu. As someone older, and perhaps wiser with life experience, Cao joked that she was the maternal grandmother of Wu and Dong Ruo-Fen. Although the Agricultural School was in the countryside away from the main campus, Cao visited the main campus frequently, and each time would bring a couple of delicious dishes to treat Wu, Dong, and Zhu Ru-Hua.[12]

Hu sometimes visited Cao in National Central University. When he inquired about his favorite student Wu during one of these visits, he found out that Cao and Wu were already friends. When Hu visited again, Cao cooked dinner and also invited Wu. They chatted after dinner, urged Hu to do calligraphy, and enjoyed each other's company.[13]

The acquaintance of Cao and Wu at National Central University was rather brief, but their friendship lasted a lifetime. Cao went to Cornell University to study the genetics of cotton in 1934, returned to China and taught in several universities. During the Sino-Japanese War, she met someone in Sichuan and was talking seriously of marriage. Unfortunately, the relatives of the groom-to-be heard of the details of the past Hu-Cao romance from Mrs. Hu Shih, and the marriage plan was cancelled. Cao was heartbroken again, and ready to convert to a Buddhist nun. Her older brother intervened, but she was depressed beyond salvation.

Cao did research on the genetics of potatoes in Shenyang Agricultural School, and succeeded in producing a high yield, high quality strain. Then the Cultural Revolution erupted; Cao was forced to return to Jixi village, lonely and sick. She visited an aunt in Shanghai in 1972, and learned from alumni of National Central that Wu was finally planning to return to China after nearly 37 years in America. She decided to stay and wait for a reunion. Wu did return in August 1973, but sadly Cao passed away in the spring of that year.[14]

Cao was an ill-fated talent. Wu was very sympathetic to her. Cao wrote a poem "After Xi-Jiang-Yue" in 1961, which had strong hint of her love for Hu Shih forty years before. In her will, she wanted to be buried by the road in "Eighth Town, Jixi", as the road leads to "Seventh Town" where Hu Shih was born. She wished Hu would visit her grave on his way home.[15]

Compared to the rocky love life of Cao, Wu was very focused in her studies, and did not get distracted by romance. In National Central University, all the girls formed a tight-knit circle, and any news of romance traveled quickly. Her close friends all said that Wu had no boyfriend as far as they knew.[16]

Wu Chien-Shiung spent four years at National Central from the age of eighteen to twenty-two. She blossomed into a "south-of-Yangtze" beauty with delicate features and a healthy body. Some girls even became obsessed with her.[17] There must have been boys falling head-over-heels for her, in secret or in public.

She was always proper, but passionate in private. As she had great love for her friends and life, she must also have yearned for romance as a young maiden.

Wu had the ambition to succeed in science very early in her life, and avoided the distraction of a serious romance. Perhaps she had channeled her desire into strong bonding with girlfriends, and the profound admiration of Hu Shih and another professor at National Central.

Wu Chien-Shiung single-mindedly focused on learning in the four years in National Central University. In the laboratory, she gradually grasped the process of establishing knowledge through an accumulation of experimental results, and the seemingly lonely life of an experimentalist. Actually, there was an exciting world community of like-minded experimental scientists out there, and Wu would soon join them exploring the secret of the universe. She learned much from the National Central's Chairman Fang Guang-Qi who taught optics, and Shi Shi-Yuan who taught modern physics.

Wu could not be totally detached from current events. Japan invaded Manchuria in the "September 18 Incident" during her freshman year, and their army landed in Song-Hu during the "January 28 Incident" the following year. Both incidents engendered much anger in the public. China had been humiliated repeatedly, from the defeat in the Sino-Chinese War in late 19th century, to the treaty of many concessions after the invasion of the "Eight-Nation Expeditionary Forces". The suffering and anger from the oppression of foreign powers and imperialism pushed the public to a breaking point. Young students held street protests almost every day, begging the government to be tough, even to declare war on Japan.

Because National Central University was located in the capital, the student movements were especially active. The protesting students occupied the university office in October 1932, and President Zhu Jia-Hua resigned. He was replaced by Luo Jia-Lun, but not until August the following year.

The political scene then was very risky and fluid. The protesting students could be suspended. Driven by the old saying "*The fate of a country rests with everybody*", there were still many students risking it all. One of the leaders was Yu Ji-Zhong — then a student in National Central and later the CEO of *Reading Times* in Taiwan. Yu was handsome, an eloquent speaker, a dedicated patriot to the students, but an activist against the government agencies.[18]

Wu was not an activist. Because she was a good student, she ran less risk of being suspended. Her father also had a record of supporting revolution, even assisting in the transportation of personnel and supplies of the 19th Army in the "Battle of January 28", and probably would not object to her participation. Wu was elected to be one of the protest leaders.

The protests and demonstrations led by Wu were rather low profile, as she would avoid areas with the State Department and the Press, and pick dates just before holidays. She figured that the protests would not last for too long, as the students were eager to return home for the holidays.[19]

Once, Wu and other students held a demonstration in the garden of the Presidential Palace in Nanjing urging the government to declare war on Japan. It was in December, and it started to snow at night. The students sat in peacefully. President Chiang Kai-Shek finally showed up very late, listened courteously to their demands, and explained, in his heavy Zhejiang accent, the difficulties and limitations of the government. The students were not totally satisfied. Chiang agreed to do what he could and urged them to go home.[20]

Wu learned from these chaotic events that one must learn to be resourceful in order to have a chance of achieving a better world. She focused her study on physics for this reason, which was shared by most of the intellectuals at that time.

President Luo Jia-Lun wrote the article "National Central University Grown up under the Bombs". He wrote: "After the 'September 18 Incident' and the 'Battle of Song-Hu', Chinese academics, especially those in universities, did not give up, but rather doubled their effort. The period from the 'January 28 Incident' to the 'July 7 Incident' was the time when Chinese Higher Education made the most rapid, most solid advancement".

In this subdued environment, her study progressed rapidly. Wu audited courses in other departments. She gained attention and praise from fellow students and professors. Physics Professors, like Shi Shi-Yuan, Fang Quang-Qi, and others, concurred that she would have a bright future beyond any limit.[21]

Wu graduated from National Central University with top honors in 1934. She did her senior thesis with Shi Shi-Yuan, and went to Zhejiang University as a Teaching Assistant for one year. Zhejiang University was an up-and-coming university attracting many young scholars, and was honored as "Cambridge of the East" by the visiting British scholar J. Needham. The Physicist Wang Kan-Chang joined Zhejiang University after his return from Europe in 1934 and a brief stay in Shandong University. Wang later contributed critically to the development of nuclear bombs in China, and was honored as the "Father of the Chinese Hydrogen Bomb". The Nobel Laureate T. D. Lee was a student of Wang at Zhejiang University.

During Wu's year there, neither Lee nor Wang had yet arrived. The chairman of the physics department, Zhang Shao-Zhong, was very nice to Wu. He asked Wu one day near the beginning of summer if she wanted to work in Academia Sinica. Academia Sinica had physics and chemistry Research Departments housed in a grand four-story building, facing the scenic Zhaofeng Park near the end of Yu-Yuan Road west of Shanghai. Usually, one had to apply and take an examination just to get admitted to work there. Wu did not find out who had made the recommendation for her to work there, but she happily took the opportunity anyway.[22]

The director of the physics department was Ting Xie-Lin, and Wu's advisor was a female Research Professor Gu Jing-Wei. Before Wu arrived, Gu ran into Cheng Chong-Dao, then working in the Chemistry Department, and inquired about Wu. Cheng said, "I know her, Wu was a towering figure among our classmates. She is intelligent, capable, serious, and good-natured." Gu was delighted that the physics department had recruited such a talent.

Gu returned to China after earning a Ph.D. from the University of Michigan. Her laboratory had two parts: a large dark room and a small meeting room. Wu and Gu, both ambitious, new-generation scientists, designed an experiment to measure the spectrum of a certain gas at low temperature, as a means to explore the inner structure of atoms. They had to set up the apparatus, purify the gas and achieve a high vacuum. They worked day and night in the dark room, at times forgetting about other necessities.[23]

Gu had teaching duty, and could go to the lab only once a week. Most of the time, it was Wu who worked there alone, somewhat like a present-day graduate student or postdoctorate.

People in Academia Sinica usually took a short break after lunch, taking a short nap on their desks, or enjoying a stroll in Zhaofeng Park. Wu frequently forgot about lunch, and would embarrassingly emerge from the dark room, rubbing her eyes to adjust to sunlight when visited by friends at lunchtime. She also would work on Sundays.[24]

Cheng Chong-Dao remembered that Wu had a certain routine. She would go to the beauty salon every Tuesday, and movie theater once a week. She appreciated fine arts, photos and paintings. Her uncle once traveled to Huang Shan (Yellow Mountain) and gave her several photos — mountain peaks (appearing like islands) in an "ocean" of clouds, along with strangely-shaped pine trees. Wu liked them a lot, and showed them off to friends. Her apartment nearby was nicely furnished, and her occasional cooking well-appreciated.[25]

Wu decided to continue her study abroad, and worked hard on her English. She was strongly encouraged by her advisor Gu, and her favorite teacher Hu Shih also visited her laboratory in Academia Sinica to show his support. Wu always felt that she benefited very much by such nurturing teachers in China.[26]

She received funding and assistance from her uncle in July or August 1936, and acceptance from the University of Michigan. Wu planned to travel to the United States with Dong Ruo-Fen, who had grown up in the same village.

This was before the jet age, and the only travel means to the US was by ship. Wu and Dong tried to buy two second-class tickets, but all cabins were sold out except one first-class cabin. They would have missed classes taking the next scheduled ship a month later.

Wu then negotiated with the salesman: "Why don't you sell us this last first-class cabin at second-class price? We promise to use only the second-class dining room." The salesman refused, but Wu convinced him to check with his supervisor.

When told of the negotiation, Wu's father thought her plan was not possible. Wu checked back the next day, and was pleasantly surprised that the supervisor had agreed to the deal! Wu set out to travel all the way across the Pacific Ocean to the US on the ocean liner *President Hoover*. She said the other reason she preferred second-class dining was that first-class passengers had to dress up for dinner every day — much too formal and uncomfortable for her.[27]

Wu left in August 1936. Her parents and relatives all gathered in Huang Pu Bund to see her off. They stayed on shore as the ocean liner was

A painting of Wu Chien-Shiung with the background of the Pacific Ocean on her way to the US.

too big to dock by the Bund, and passengers had to be transferred on board by small boats.

She remembered that her mother cried a lot that day, and her beloved father and uncle were sad to see her leave. Wu originally planned to study abroad for just a few years, and return to China. But the trip turned out to be a 37-year ordeal, and she would never see her parents again.

It was 1936. The "Marco Polo Bridge Incident" the next year ignited an all out Sino–Japanese War that would last eight long years.

Notes

1. *Seventy Years in National Central University*, National Central University, 1985, Zhong-Li, Taiwan.
2. Based on the memoirs of Sun Duo-Ci, Wu Zi-Wo, and Cheng Chong-Dao, all classmates in National Central University.
3. Based on *Physics of the Twentieth Century*, C. N. Yang; and *From X-Rays to Quarks*, Emilio Segrè, San Francisco, 1980.
4. Based on interview with Wu, memories of her classmates, and *From X-Rays to Quarks*, Emilio Segrè, San Francisco, 1980.

5. *Dr. Wu Chien-Shiung and Scientific Research*, Wu Zi-Wo, in *Tribute to Dr. Wu Chien-Shiung on the occasion of receiving the Special Contribution Award from Chi-Tsin Culture Foundation*, page 18, July 1965, Taipei.

6. *Wu Chien-Shiung — A Chinese Madame Curie*, Sun Duo-Ci, in *Tribute to Dr. Wu Chien-Shiung on the occasion of receiving the Special Contribution Award from Chi-Tsin Culture Foundation*, page 20, July 1965, Taipei.

7. Interview with Zhang Xi-Ying, August 15, 1989, London residence. Also in the memoir of Cheng Chong-Dao.

8. Interview with Zhang Xi-Ying, August 15, 1989, London residence.

9. Interview with Zhang Xi-Ying, August 15, 1989, London residence.

10. Interview with Yen Mei-He, June 13, 1990, New York City.

11. *Hu Shih in Love*, Fu Jian-Zhong, *Reading Times Weekly*, Foreign Edition, page 31, July 1987. It quoted *Casual Hu Shih*, Shi Yuan-Gao, China.

12. Interview with Wu Chien-Shiung, August 21, 1990, New York residence.

13. Interview with Wu Chien-Shiung, December 5, 1989, New York residence.

14. *Hu Shih in Love*, Fu Jian-Zhong. Also interview with Wu Chien-Shiung, August 21, 1990, New York residence.

15. *A Visit to the Hometown of Hu Shih*, Li You-Ning, *Reading Times Weekly*, Foreign Edition, page 69, Number 262, March 3, 1990.

16. Interview with Zhang Xi-Ying, August 15, 1989, London residence.

17. Based on the memoirs of Sun Duo-Ci, Wu Zi-Wo, and Cheng Chong-Dao.

18. Interview with Zhang Xi-Ying, August 15, 1989, London residence.

19. Interview with Wu Chien-Shiung, January 6, 1990, New York residence.

20. Interview with Wu Chien-Shiung, September 25, 1990, New York residence. Interview with Yen Mei-He, June 13, 1990, New York City.

21. Based on the memoirs of Sun Duo-Ci, Yen Mei-He, and Cheng Chong-Dao.

22. Interview with Wu Chien-Shiung, December 5, 1989, New York residence. Based on the memoirs of Cheng Chong-Dao.

23. *Wu Chien-Shiung in My Heart*, Cheng Chong-Dao, in *Tribute to Dr. Wu Chien-Shiung on the occasion of receiving the Special Contribution Award from Chi-Tsin Culture Foundation*, page 23, July 1965, Taipei.

24. Interview with Wu Chien-Shiung, December 5, 1989, New York residence. Based on the memoirs of Cheng Chong-Dao.

25. *Wu Chien-Shiung in My Heart*, Cheng Chong-Dao, in *Tribute to Dr. Wu Chien-Shiung on the occasion of receiving the Special Contribution Award from Chi-Tsin Culture Foundation*, page 23, July 1965, Taipei.

26. Interview with Wu Chien-Shiung, December 5, 1989, New York residence.

27. Interview with Wu Chien-Shiung, September 13, 1989, New York residence.

Chapter 4

A Rising Star in Berkeley

When a twenty-four-year-old Wu Chien-Shiung arrived in San Francisco, California in August 1936, she had only a vague impression of the US. Her teachers Hu Shih and Gu Jing-Wei had studied in the US, and she had learned a little bit about its culture from newspaper and magazine articles. Like a typical Chinese intellectual, she had a good first impression of this land. After all, a country named "Beautiful Country" in Chinese should not be bad.

Everything that happened on the ocean liner, or during the crossing of the Pacific itself, was all fresh and interesting. Wu was passionate about life and the future, and she sucked up the new experiences like a sponge.

The ocean liner, *President Hoover*, made a stop in Japan. Wu had no good feeling about this neighbor, as she was haunted by the bitter memories of Japanese oppression in China.

The wide-open US was then very different. In 1936, it was rising as a world power in the 20th century. Many immigrants, including many first-rate intellectuals, were driven out of their European countries with the rise of fascism. They helped to build this country.

Wu's first stop, San Francisco, was the new frontier on the West Coast of this young country. A youthful Wu just got caught up in the excitement.

She was supposed to stay in San Francisco for just a week, visiting a classmate, Lin, whose husband, Guo, was a professor at UC Berkeley. She planned to continue her journey eastward to the University of Michigan. This one-week visit turned into a multi-year stay at UC Berkeley.

School had already started at UC Berkeley when Wu arrived. The President of the Chinese Students' Association, Victor Yang, was very active and enthusiastic. He told Wu that a Chinese graduate student in Physics, Luke Yuan, had arrived just two weeks before, and could show her

Luke C. L. Yuan (*extreme right*) with classmates at Yenching University in 1933.

around. Luke was the grandson of Yuan Shih-Kai, a cabinet member in the Qing Dynasty, and later a powerful figure in the early days of the Republic of China. Wu did not know of this relationship.

Wu visited the hilly campus of UC Berkeley with Luke as a guide. She had investigated the spectra in the X-ray crystallography experiments at Academia Sinica, and had developed a first-hand appreciation of the experimental setup. She could not restrain her excitement of seeing the many experimental apparatuses in the physics department.

What impressed Wu most was the Radiation Laboratory in Le Conte Hall, built by Ernest Lawrence. This lab later expanded into the Lawrence Berkeley Laboratory. It had a 37-inch cyclotron — the first in the world, able to accelerate charged particles to bombard and break up different nuclei. Wu admired this hottest apparatus in nuclear experiments.[1]

The physics department at Berkeley did not have the long proud record of its counterparts on the East Coast: Harvard, Yale, and Columbia. But it attracted many young and top scientists. Lawrence, who invented and built the cyclotron, was only 35 years old. There was also this absolutely brilliant, young theoretical physicist, J. Robert Oppenheimer.

Oppenheimer had graduated from Harvard in chemistry, and started touring Germany and England in 1925. It was the dawn of quantum mechanics in Europe, and he became close friends with top physicists like Max Born and Paul Dirac. His first-rate talent in physics, sharp tongue, insatiable curiosity, and strong ego all made him a legend on campus after he joined the physics department as a professor in 1929. He was then 28 years old.

Wu was brilliant, ambitious, strong-willed, and had great taste in research. She found the physics department at UC Berkeley irresistible, a dream place to search for scientific knowledge. She quickly decided not to go to Michigan and to stay at Berkeley.

There was another factor in her decision.

Not only was Berkeley a hub of talents, but California was then more liberal. The East Coast, and more so Michigan in the Midwest, was fairly conservative.

When visiting Berkeley, Wu was alarmed to learn that a new student center in the University of Michigan, built with donations from students (many of them females), did *not* allow females to use the front entrance.

Wu was surprised by such acts of discriminations against women. She and other female students had never been treated unfairly in China. This was also the first occasion when she had a different opinion of the society in the US.[2]

How could she go to a place where she could be treated as a second-class citizen? In addition, the University of Michigan had more than 600 Chinese students; Wu would not like to go all the way to the US and end up socializing only with Chinese groups.[3]

Accompanied by Luke, Wu went to see the chairman of the physics department, R. Birge. Birge had been criticized as prickly, narrow-minded, and prejudiced against foreigners, particularly Chinese, women, and anyone with a foreign accent. But he did make critical contributions to developing the Berkeley physics department into a world-class institute during his 20-year tenure running the department (1933–1955).[4]

If Birge had any prejudice against the Chinese or women, it did not show, or stop him from recognizing Wu's unusual talent in physics. Birge understood that it takes first-rate students, as well as first-rate professors, to build up a good department. He made an exception to accept Wu into the graduate school even though the academic year had already started.

All the while, Wu's travel companion Dong Ruo-Fen was both surprised at and resentful of her change of mind. Dong went on to the
University of Michigan alone, studied chemistry, returned to China, but
came back to the US many years afterwards. They were no longer dear
friends as before.

Wu stayed in Berkeley, and moved into the International House nearby.
She worked hard learning English and understanding the US culture in
addition to science. But Wu maintained two things Chinese: dresses and
food. She always wore traditional Chinese gowns — the high-collared
qipao.

She could not stand the Western food served in the cafeteria the very
first morning in the International House. That afternoon, she was happy to
find a small shop nearby serving spring rolls and tea (more to her taste).

Wu also met a German student called Ursula Schaefer on her first day
in Berkeley. Ursula was pure German, but liked everything Chinese,
including food. She could not stand the cafeteria food in the International
House either.

Wu found a solution soon. With some help from a friend, she discovered a Chinese restaurant, the Tea Garden. The owner was very nice and
made a deal with Wu: she and her friends could have dinner (with four
dishes, soup, and all-you-can-eat rice) there for 25 cents per person. The
only restriction was that they could not order from the menu — the restaurant would serve whatever "surplus" they had that day. Wu would go there
for dinner, frequently with friends like Luke Yuan, Ursula, and Willis Lamb
(who would later win a Nobel Prize, and marry Ursula). It was indeed a
good deal for four hungry graduate students to eat for only a dollar.[5]

Soon after Wu settled down, her favorite teacher, Hu Shih, visited his
good friend Professor Ma in Berkeley. Knowing that Wu was a former student of Hu, and that Luke's father, Yuan Ke-Wen, was an acquaintance of
Hu in Shanghai, Professor Ma invited both Wu and Luke to dinner with
Hu Shih at his house.[6]

Wu and Hu had a long conversation that evening. Hu was very happy
to learn of her determination and persistence. He wrote her a long letter the
following day offering much encouragement and citing his expectation.
After Hu passed away many years later, the letter was returned to Mrs. Hu,
and it is now preserved in the Hu Shih Memorial in Nankang.

Wu arriving in Berkeley in 1936.

Studying in Berkeley.

At UC Berkeley in 1938.

Wu in Berkeley.

With a foreign friend.

Her uncle Wu Zhuo-Zhi supported Wu during the first year. One can tell from her dressing in the photos then that her family must have been quite well-to-do. On the other hand, Luke was frugal in his spending as he depended solely on his scholarship for tuition, room and board. Ursula was exiled from Germany because she disagreed with the policy of forcing out the Jews there. Her family could no longer send her money, after her passport was canceled. The twenty-five-cents meal was her life-saving deal.[7]

Wu was in a class of about 15 students in the first two years. Some of her classmates did very well afterwards. For example, Robert Wilson became the founding director of the Fermi National Accelerator Laboratory, a high energy physics research center near Chicago. He even designed the signature office tower there. He retired and lived with his artist wife near the scenic campus of Cornell University. When I interviewed them in September 1989, I could no longer see the image of a strong man running Fermilab.

The other one was G. Volkoff who became the President of the Canadian Physical Society, and Science Advisor to the Canadian Government. He contributed greatly to the development of physics in Canada. These former classmates all praised her talent and persistence in the student days, and her lasting achievements in physics.

In addition to her scientific talent and persistence, Wu's professors and classmates were also impressed by her absolute determination to succeed, and her sensitivity to achievements.[8]

Volkoff recalled that her English was not that proficient, and she sometimes had difficulty taking notes in the class without missing something. Volkoff might not be a good physicist (so he said), but he was certainly a good student and note-taker. Wu (and also Luke) would borrow and copy his lecture notes.

Volkoff was outgoing and an eloquent speaker. He had a special feeling for China, as he had lived in Harbin for three years as a high school student. He would lend support to Wu and Luke, and was of course a member of the twenty-five-cents dinner club.[9]

Lamb recalled taking a course in quantum mechanics with Wu, Wilson, and Volkoff. The professor was none other than Oppenheimer, who knew the subject inside out. He attracted a group of the most brilliant students, who were totally mesmerized by his talent and charisma.

Wu was one of the admirers. Many years later she still referred to him affectionately as "Oppie", and as someone absolutely brilliant.[10] In return, Oppenheimer praised highly her experimental work. They maintained this close teacher-student bonding. When he did not select her for a research project, Wu was disappointed and in tears from anger.[11]

Wu finished the first year with good grades. She applied for a scholarship to free herself of the financial support from her uncle. At the time the US discriminated against Asians. The chairman of the physics department, Birge, did not want to raise the ire of the Board of Trustees, and agreed to award Wu and Luke readerships with smaller stipends.[12]

Luke Yuan was not that content with the situation, and applied to the California Institute of Technology near Los Angeles. The President, Robert Millikan, personally sent Luke a telegraph, offering him a scholarship, and asked for an immediate acceptance. Luke accepted the opportunity, but in doing so made Birge unhappy.[13]

Wu continued her serious study in Berkeley. She attained the status of a star, as she was talented, hard working, pretty, elegant, and also outgoing. There were many admirers of this beauty in the physics department; some jokingly included the sound of her last name "Wu" (sound of longing) in a love song circulating around the campus.[14]

Wu stood out in many ways. Friends in the physics department and the International House could not help admiring this slender beauty wearing a high-collared *qipao* with slits on both sides. Some called her Miss. Wu, some Chien-Shiung, and close friends affectionately called her "Gee Gee." Her friends believed that the last nickname was the foreign pronunciation of "Zi Zi" ("elder sister" in Chinese).[15]

Gee Gee made friends with others in the International House. Adina Wiens, an American who studied Nutrition Science, moved there in 1937. They became dear friends. They were joined by an overseas Chinese called Eda. Adina recalled that the threesome fitted each other like a pair of "worn shoes".

There was also Margaret Lewis, an American who had earned a Ph.D. in physics from Johns Hopkins University, and had come to Berkeley as a postdoc in 1937. She and Wu became life-long friends. Wu always worked hard and rarely participated in recreation. Margaret once asked Wu to go to Yosemite National Park, but Wu turned her down because she had to study.

The two went to Lake Tahoe in the summer of 1941, but Wu would study in the morning.

Wu, Adina, Eda (the three "worn shoes"), Margaret, and other friends were all frequent members of the twenty-five-cents dinner club.[16]

Wu was proper and serious in social functions. Her friend Ursula quickly discovered a rational and honest personality beneath her charming and proper exterior. Wu was always truthful to her friends, and would not even tell a white lie. If she disagreed with Ursula's behavior, she would frankly tell her so.

Wu once gently scolded Ursula. Ursula was embarrassed for many days, as no one else would dare to do that to her face. Such honest advice actually nurtured their friendship. Wu would give her advice with a smile, and often cover her mouth with her hand while smiling. Ursula said that she might not have ever seen her teeth.[17]

Wu was truthful to her friends, and expected the same moral standard from them. Margaret recalled that she had accidentally hurt Wu's feelings once:

It was a Saturday evening and Wu was in Margaret's room chatting. Margaret got a phone call inviting her to a dance. To an American girl, it was important. Margaret went to the dance and left Wu alone. She found out later that Wu was very offended.[18]

Wu usually did not like going to dances or other parties. She would sometimes go to concerts or operas with Adina, Margaret and Ursula, or bring these "foreigners" to Beijing operas.

Margaret recalled a special gathering they both attended, but they swapped dresses before going — Wu in a Western dress and Margaret in a *qipao*. Everybody enjoyed the stunt.[19] Wu was never crazy for the performing arts, and went out less and less often as her workload increased.

It became obvious that scientific research was always the top priority in Wu's mind. Wu knew that she possessed great talent, ample interests, and strong determination. All her old friends said that Wu was so ambitious that she would give up anything for research.[20]

After completing her course work in the first two years, Wu got ready for her Ph.D. thesis work. She had to find an advisor, select a research topic, and earn the credentials to be admitted to long-term experimental work.

▲
◀

Wu and Margaret Lewis.

Wu's unusual talent for physics included a solid foundation in X-ray crystallography experiments in Academia Sinica. The great Ernest Lawrence, Director of the Radiation Laboratory, became her official advisor.

Her actual advisor was Emilio Segrè. Segrè worked on nuclear experiments with Enrico Fermi in the 1920s in Rome. Many famous physicists had come from the research group led by Fermi in Rome, including Edoardo Amaldi and Bruno Pontecorvo, who gave away the atomic bomb secret to Russia and was exiled there afterwards. They passed away in 1989 and 1993, respectively.

Segrè was working at the University of Palermo in Sicily, Italy. He visited New York, Chicago, and Berkeley in July 1938, and planned to return home in October. But he learned that Mussolini in Italy might adopt the same anti-Semitic policies of Hitler, and he did have a trace of Jewish blood. Segrè decided not to return to Italy and stayed in Berkeley, a far cry from Europe.

His earlier experiments and his research in Europe had already made Segrè a rather prominent figure in physics. Although he held no official position in Berkeley, and was paid by the Rockefeller Foundation at first,[21] he had many collaborators right away.

There was a postdoc, Alexander Langsdorf Jr., working for Segrè. Langsdorf was a Jew. Although there was no outright oppression of Jews as in Germany, the US did discriminate against them and had a strict quota for professorships in universities. So it was still difficult for Jews to find jobs.

One day, a happy Langsdorf told Wu that he had found a tenured position in St. Louis and would move there. He also said that Segrè would have openings, and invited Wu to their group meeting the following week. Wu went, became very interested in the work, and joined Segre's research group.[22]

Wu formally started her life-long pursuit of experimental nuclear physics in 1938. The field of study was blossoming, building on the foundation of the many developments in the early 20th century.

Lawrence directed her first experiment. The topic was to investigate two modes of X-rays excited by electrons emitted in the beta decay of radioactive lead. This experiment was directly relevant to the theoretical work of George Uhlenbeck, Hans Bethe, and Walter Heitler.

Nothing shook the world of nuclear physics more than the discovery of uranium fission near the end of 1938. The discovery was published in

Wu (*second from right*) in Berkeley with J. Robert Oppenheimer (*third from right*), her advisor, Emilio Segrè (*fourth from right*) and good friend Margaret Lewis (*extreme left*).

Science in January 1939. Every scientist in this field rushed to start relevant experiments.

Directed by Segrè in 1939, Wu worked on experiments that studied the products of uranium fission. This series of experiments, advised by Segrè, but mostly done by Wu, had important consequences. One of the results turned out to be critical later in the development of the atomic bomb — in the Manhattan Project in the US.

From the very beginning Wu possessed a research style that featured relentless precision and accuracy. Both her nominal advisor, Lawrence, and collaborator, Segrè, highly praised her work ethic.

In the biography published after his death, Segre described his first student in Berkeley, Wu, as crazy about physics, almost obsessed, talented and very brilliant.[23]

Segrè also described Wu as very pretty and elegant in a Chinese *qipao*. There were many admirers following her around like a queen on campus. Segrè admired her and liked her.[24] Their teacher-student relationship was as close as a father-daughter one. This life-long relationship was maintained despite an incident in which Segrè publicly complained that Wu was too demanding of her students — almost like a slave driver.[25] Wu was not pleased by this criticism.

In research, Wu had the same determination and persistence in those early days. Glenn Seaborg was most impressed by her stubbornness and determination. He went to Berkeley two years before Wu, earned a Ph.D. three years ahead, and won a Nobel Prize in 1951 with the discovery of transuranium elements.

Seaborg was invited by Lee Yuan-Tseh to visit Taiwan in 1989. He was tall. In his office in the Lawrence Berkeley Laboratory, there were photos of him and US presidents — Nixon, Ford, and Carter, as well as celebrities like Ann-Margaret. He kept a diary, from the Berkeley days to his Taiwan visit.

Sitting in a tall chair in March 1990, Seaborg recalled that Wu was one of the few female students in Berkeley, and that she was very bright. As all the students were bright, her additional determination and persistence left him with a lasting impression.

Seaborg recalled that he and Wu both used the cyclotron. Wu could be very tough when negotiating a better beam position for her experiment. Her style was polite but firm.[26]

Totally devoted to her experiments, Wu would usually stay in her laboratory until very late at night. The physics department was concerned about her safety walking back to the International House at midnight, and arranged for another night-owl student, Robert Wilson, to drive her home in his old car. At three or four in the morning, Wilson would stop by her lab, saying: "Miss Wu, it's time for you to go home."[27]

Wilson was a talented experimentalist, and one of Wu's admirers (secretly courting her). He worked hard at night in his lab, and worked back home in Ohio every summer to support his studies. Wu was impressed, and she appreciated the independent spirit of American students.[28]

Just like any successful scientist, Wu idolized world-class scientists and hoped for opportunities to learn from them. In the Berkeley years, the great Danish scientist Niels Bohr (who proposed the atomic model) visited the Berkeley campus. A reception was held in the International House but attendance was limited to professors only. Being a student, Wu was disappointed to tears because she was not allowed to attend.

Wilson had an idea. When Bohr was standing alone briefly, he grabbed Wu, moved towards Bohr, and introduced Wu and himself. Bohr was fascinated by such a smart young female student, and chatted with them while

The host of an award ceremony at UC Berkeley escorting Wu.

enjoying some tobacco that Wilson had brought along. Bohr's English was poor, Wu and Wilson did not get much insight from him, but were excited by the experience nevertheless.[29]

There was another towering figure in physics — Wolfgang Pauli. When visiting the US he lectured in Berkeley and Segrè invited Pauli for dinner and asked him if he wanted to see others. Pauli suggested a Chinese student and Segrè naturally asked Wu.

Pauli once translated a German book on the *I Ching*. He talked to Wu a great deal during dinner, believing that the Chinese had a profound understanding of the universe.[30] Pauli was a genius, famous for his sharp tongue and inquisitiveness. He caught people's attention as a student by saying: "Einstein's lecture was not so stupid."[31] Pauli's accomplishments in physics were seminal. He praised Wu's work highly, and they developed a close friendship ever since.

Wu obtained her Ph.D. in 1940. Her thesis had two research topics, as discussed before. Both of the experimental results were published in the *Physical Review* and had major impact. This Ph.D. research was miles ahead of the average thesis.

Wu remained in Berkeley for two years as a postdoc. She investigated the fission products of uranium and the radioactive isotopes of some elements, frequently using the neutron beam produced in the cyclotron. Lawrence was promoting radiological treatment of cancer using the cyclotron, partly to offset its expenses. Wu would go there at six in the morning, and start her experiment as soon as the treatments of cancer patients were done for the day.[32]

She had other collaborators besides Segrè during this period. According to Seaborg's diary, Wu worked with G. Friedlander (a student of Segrè) and C. Helmholz to investigate a radioactive isotope of mercury using the neutron beam produced in the cyclotron, with heavy helium as a target.

At the annual meeting of the American Physical Society held at Stanford University on December 19, 1941, Wu and Segrè presented their experimental results on the artificial radioactivity of some rare-earth elements.[33]

Segrè worked closely with Wu and they had frequent discussions. They were toying with an experiment to find out whether the half-life of beryllium would change with a modification of the electron number. They did not know then that beryllium was poisonous. Fortunately, they could not get a pure enough sample, and had to stop because of the onset of the war, thus surviving a possible disaster.[34]

The distinguished results on nuclear fission and radioactive isotope established Wu as the "authority" in the minds of Oppenheimer and others. She was often invited to speak at conferences. Segrè would even borrow her lecture notes, when invited to give seminars on nuclear fission.[35]

There was another time when Segrè showed Wu a sample irradiated by a neutron beam. Wu did some measurement and identified it correctly as Ru-45. Before he passed away in 1989, Segrè left this same sample in Berkeley.[36]

Wu was focused on her research, but she never forgot her purpose for studying abroad. After the Sino-Japanese War started, she struggled with the idea of whether she should return to China. Then Japan started the Pacific War, and the postal service between the US and China was cut off. Wu was left to worry helplessly about her hometown and family.

Perhaps because of her Chinese roots, Wu would unconsciously speak Chinese with good friends. Her English, with a heavy Shanghainese accent, was difficult to follow, and she confused the word "he" for "she" at times. But her written English was more fluent and elegant. One famous story relates that once she became so involved in giving a seminar that she unconsciously started writing a physics formula from right to left, like Chinese.

Wu had unquestionably became a legend in Berkeley. Scientists of her professors' generation praised her, her classmates and collaborators agreed that she would have a great future. Her reputation started to spread outside her academic circle and even prompted an article in the local newspaper. The news of a "Chinese Madame Curie" also began to circulate in China.

Wu Chien-Shiung had become a shining star of science.

Notes

1. Interview with Wu Chien-Shiung, August 28, 1990, New York residence.
2. Interview with Wu Chien-Shiung, August 28 and October 26, 1990, New York residence.
3. Interview with Wu Chien-Shiung, August 28, 1990, New York residence. *Nobel Prize Women in Science*, S. B. McGrayne, Carol Publishing Group, New York, 1993.
4. *A Mind Always in Motion: The Autobiography of Emilio Segrè*, Emilio Segrè, UC Berkeley Press, 1993.
5. Interview with Wu Chien-Shiung, September 15, 1990, New York residence. Interviews with George Volkoff and Robert Wilson.
6. Interview with Wu Chien-Shiung, September 13, 1989, New York residence.
7. Personal letter to the author from Ursula Schaefer, March 21, 1994.
8. Interview with R. Wilson, September 21, 1989, Ithaca residence and interview with G. Volkoff, October 12, 1989, Faculty Club, University of British Columbia, Vancouver.
9. Interview with G. Volkoff, October 12, 1989, Faculty Club, University of British Columbia, Vancouver.

10. Interview with Wu Chien-Shiung, July 31, 1990, New York residence.
11. Interview with Adina Wiens, October 16, 1989, San Francisco residence.
    Interview with Xu Jing-Yi, October 16, 1989, San Francisco residence.
12. Interview with Luke Yuan, September 8, 1989, New York residence.
13. Interview with Luke Yuan, September 8, 1989, New York residence.
14. Interviews with George Volkoff and Robert Wilson. Also *Nobel Prize Women in Science*, S. B. McGrayne, Carol Publishing Group, New York, 1993.
15. Wu Chien-Shiung and her friends thought that "Gee Gee" must be a phonetic variation of "Zi Zi".
16. Interview with Adina Wiens, October 16, 1989, San Francisco residence. Interview with Margaret Lewis, February 22, 1990, Library of Harvard University Observatory, Boston.
17. Interview with Wu Chien-Shiung, July 31, 1990, New York residence. Personal letter to the author from Ursula Schaefer, March 28, 1994.
18. Interview with Margaret Lewis, February 22, 1990, Library of Harvard University Observatory, Boston.
19. Same as 18.
20. Interview with R. Wilson and others.
21. *Lawrence and His Laboratory*, J. L. Heibron and R.W. Seidel, UC Berkeley Press, 1989.
22. Interview with Wu Chien-Shiung, August 28, 1990, New York residence.
23. *A Mind Always in Motion: The Autobiography of Emilio Segrè*, Emilio Segrè, UC Berkeley Press, 1993.
24. Same as 23.
25. *Newsweek*, May 20, 1963.
26. Interview with Glenn Seaborg, March 28, 1990, Lawrence Berkeley Laboratory.
27. *Nobel Prize Women in Science*, S. B. McGrayne, Carol Publishing Group, New York, 1993.
28. Interview with Wu Chien-Shiung, September 20, 1989, New York residence.
29. Interview with R. Wilson, September 21, 1989, Ithaca residence.
30. Interview with Wu Chien-Shiung, January 27, 1990, New York residence.
31. *The Tenth Dimension*, Jeremy Bernstein, McGraw-Hill, New York, 1989.
32. Interview with Wu Chien-Shiung, August 28, 1990, New York residence.
33. *Diary of Glenn Seaborg*, March 28, 1990.
34. Interview with Wu Chien-Shiung, February 27, 1990, New York residence.
35. Interview with Wu Chien-Shiung, September 18, 1989, New York residence.
36. Interview with Wu Chien-Shiung, September 12, 1990, New York residence.

Chapter 5

Youth and Love

Teenagers are passionate and romantic. They face different events and make different choices. These choices of youth often change their course later in life.

The young Wu Chien-Shiung was not a reckless girl. She had her burning passions beneath her proper exterior. She had imagination and high expectations in life, for her future, for herself and for love.

In 1930 Wu matriculated at the National Central University, when she was eighteen years old. It was a blossoming time for any late teen, and she was radiant even in her modest dress. She was a beauty and a talent by all accounts.

A classmate at the National Central University, the artist Sun Duo-Ci wrote an article with the following description of Wu: "We were classmates in 1931.... Chien-Shiung then was petite, active, and agile. She was from Taicang, Jiangsu. Wu had a pair of expressive eyes, a delicate mouth, and short hair. She wore flats, and modest but well-tailored *qipao*. She stood out among some 200 female students. She had many male admirers; even some girls were obsessed with her."[1]

Wu knew that she was an attractive belle, but she refrained from the distraction of romance. The National Central University was relatively small, and any romance could hardly be kept secret. Her classmates all said that they were not aware of any romance involving Wu.[2]

This does not mean that Wu was not longing for love. In fact, in the transitional year between Soochow Girls' High School and the National Central University, her passion might have been ignited by, and transformed into admiration of, a young scholar Hu Shih.

The admiration of Wu towards Hu, and the praise and nurturing from Hu in return, never crossed the line of a teacher-student relationship. Judging by the life experience of a seventeen-year-old girl, and from the examination of their letters, the initial teacher-student friendship unquestionably awakened her first love, and greatly influenced her later choice in love and marriage.

The teacher-student relationship started at the National China College, when she took the course "History of Ideologies in the 300-Year Qing Dynasty" by Hu Shih. Hu had a joint appointment as the College President. He traveled from Beijing to Shanghai once a week to deliver a two-hour lecture.

Hu was 39 years old then, had a Ph.D. from the US, was charismatic, and at the peak of his career. He had broad knowledge, as demonstrated by the many cases and examples he could give in his lectures. He was an eloquent orator, and he could present logical and fluent arguments. Wu was always keen to learn, and she must have found him very attractive and irresistible.

Sixty years later, in New York City, Wu still remembered fondly these lectures. Yen Mei-He, who had known Wu since their Soochow Girls' High School days, and retired from Beijing Normal University as a chemistry professor, recalled that Wu had told her personally: "Hu Shih's lectures were just great!"[3]

If a student admires her teacher, it did not allow that Wu had crossed the line. There were rumors and gossip about their relationship, their correspondence, and the special care they had for each other. Perhaps it is a natural tendency that room for imagination is created when two talented people are put together. There was even a novel *The Second Handshake*, published in China, telling the story of a famous scientist and a well-known scholar — an allusion to Wu and Hu.

The novel did not have much factual basis. In their friendship, they did correspond, Hu did visit Wu in Academia Sinica and in the US, and he did take good care of her. But Hu was married, and disciplined. He never showed any affection beyond the friendship between a teacher and his student. Wu always behaved properly as a student. She might unconsciously show a trace of love, given that she was just an ignorant child with limited life experience.

Wu was not willing to discuss her private feelings. We have to examine the letters exchanged between them to learn more of this friendship.

Hu wrote many letters to Wu, but she did not want to make them public. There are four of Wu's letters in *Selected Correspondences of Hu Shih*. As there was no other reference, we have no way of finding out whether other letters exist. These four letters all start with "Teacher Shih-Zhi" (Hu Shih-Zhi was an alias of Hu Shih) and end with "Respectfully, Chien-Shiung," and the language in two of these letters is very interesting.

One letter is dated April 3, 1940, right after Wu obtained her Ph.D. at UC Berkeley. It begins:

I received the letter you sent me from Los Angeles.

On the day you were scheduled to leave, I got up early and saw this pouring rain and I did not know whether I should be worried or thankful. When I called your place around ten in the morning, they said you had already left.[4]

Another letter, dated the night of February 24, 1941, begins:

Teacher Shih-Zhi,

I just returned from saying goodbye to you over the phone. Thinking that you have to rush on the road early in the morning, wouldn't it be nice if I could drive alone to the airport at dawn, and wish you farewell? But I know that it is not possible, I just pray here in silence and wish you a safe journey.

I have not written you lately, as I know that you are very busy. But I also do not want you to feel uneasy, thinking that I have misunderstood your motive. Please be assured that it will not be possible for any misunderstanding, given my profound admiration for you. I remember hearing the story of the misunderstanding between you and a female teaching assistant after my trip to the East Coast last year. I felt so sorry for you. Why are people so messy? Why do people like to spread gossip? Being sensitive to your grave responsibility, I don't even dare to mention my admiration. Hope you will excuse my silence. I would be very happy knowing that you have my sincere adoration.[5]

The true love story of Wu started in Berkeley, after her arrival by ocean liner to San Francisco from Shanghai, China in 1936.

She planned to stay in San Francisco for only one week visiting a class-mate named Lin. Lin's husband, Guo, was a Visiting Professor of Literature at UC Berkeley.

An American-born Chinese, Victor Yang, was the President of the Chinese Students Association. Yang was very enthusiastic. He did not fin-ish school but went to Hollywood to act in a series of movies as the second child of Charles Chan. Many years later, Wu recalled that she had advised Yang to finish school first, when told that he was dropping out to pursue his acting career. Her reason was that Americans were discriminating against the Chinese then, and would give Yang only a minor role in a movie. Wu said: "A very nice, smart, and competent guy should finish like that!"[6]

Yang turned out to be the matchmaker of Wu's eventually happy mar-riage. The newly arrived Wu was introduced to Victor, who mentioned: "A graduate student in physics, Luke Yuan, arrived from China just two weeks ago. He could show you around the physics department."[7]

Yang could not find Luke until that afternoon, and then brought him to meet Wu at the Lin's residence. It was the first meeting between Wu and Luke, in August 1936. The tour of the physics department changed Wu's plan to continue her travel east to Michigan. She stayed in Berkeley and became Luke's classmate.

Luke had arrived in San Francisco three weeks earlier. He had only US$40 plus a future scholarship to his credit. He traveled by ship for 16 days from Tianjin in a third-class cabin. The voyage was rocky. During the voyage, Luke could not stand the meals served in third class, consisting mostly of smelly fish, but did not want to spend money on the dollar-a-bowl porridge. He lost 20 pounds in 16 days.[8]

Although Luke was born in a ruling-class family, his immediate family was just middle-class. His father Yuan Ke-Wen was not in the primary branch (he was the offspring of Yuan Shih-Kai and his concubine). He wrote a poem — "Pity that there are frequent storms up there; It is not wise to climb all the way to the top floor of a mansion." — a mild ridicule of his father Yuan Shih-Kai reverting China to an empire. Ke-Wen was house-arrested. In contrast to his ambitious older brother Yuan Ke-Ding, Ke-Wen would stay away from Beijing, and socialized with intellectuals in Tianjin and Shanghai, while his wife lived quietly in Anyang village, Henan, bring-ing up Luke, his two brothers, and a sister.

Wu newly arriving in Berkeley.

Luke recalled that they had to make a trip to Beijing to greet uncle Ke-Ding every New Year's Day. Ke-Ding was very old-fashioned: all the kids were required to dress up properly, and kowtowed to him. Luke was fairly scared of this uncle, and grew to dislike Beijing. But he was obliged to make the trip every year as his family depended on the financial support from his uncle.[9]

Luke started school in Anyang. At 13, he went to Nankai High School in Tianjin for a month, but then transferred to The Academy of Modern Learning, run by a London-based missionary. Luke received a rather good science education there, with a Cambridge University educated Dr. Hart teaching Physics, and his maternal uncle teaching Mathematics. Luke matriculated at the College of Industry and Commerce in 1928 as a major in Engineering.

He transferred to Yenching University in 1930, where the Chinese theoretical physicist Xie Yu-Ming was a professor. Luke's interest in radio led to radio communications as a serious hobby. He graduated in 1932, stayed in the graduate school for two more years, and received a Master's degree.

After graduation Luke worked in Tangshan Coal Mine for one year. The President of Yenching University and later US Ambassador to China, Leighton Stuart, was also a radio communications hobbyist, and befriended Luke. Stuart knew of the scholarship at UC Berkeley and asked Luke if he was interested. This was the trigger for Luke's studying abroad in 1936.

Due to his upbringing Luke was diligent and frugal. His scholarship from the International House carried free tuition, and room and board. He was comfortable even with little money. He was polite, helpful, and always ready to volunteer to fix things around. His personality earned much respect from the Director of the International House, and also left a lasting impression on his classmate Wu.[10]

Wu had outstanding beauty and talent, and naturally attracted many admirers in the physics graduate school. In addition to Luke, there were at least two classmates courting her. One was Robert Wilson, the Founding Director of Fermilab.[11]

Wilson recalled that Wu was very charming but formal in her high-collared *qipao*. She attracted almost everyone living in the International House.[12]

Wu showed gentle charm and femininity in her dress and her countenance. But her personality was very outgoing and unpretentious. Her manner stood out, especially for a Chinese female in the 1930s. Luke recalled that Wu never thought twice about studying with him alone and late at night in the library.[13]

Her life as a graduate student at Berkeley was rather uneventful. She studied seriously and was friendly with classmates. The letters which the young Wu wrote to her dear friend Adina revealed her love of life and her friends.

Wu finished her Ph.D. and started her postdoc research at UC Berkeley in 1941. She made a trip to the East Coast in May.

From the time of her arrival in the US in 1936, Wu immersed herself totally in study and experiments, and she had little exposure to other regions in the US. Her advisor, the physicist E. O. Lawrence, suggested that she tour the East Coast, to observe the scenery and culture of the different regions along the way.

With the assistance of Lawrence and her friend Margaret, Wu took the train across the continent. She left San Francisco in April, crossed the Rocky Mountains and the Grand Canyon, and visited Chicago and

St. Louis. She arrived in Washington, DC in May, then New York City and Boston. Along the way, Wu visited many universities and met famous physicists, both of which greatly enriched her knowledge and broadened her viewpoints. She also visited many ancient sites, and learned more about traditions and history of the US.

Wu attended the annual meeting of the American Physical Society in Washington, DC in May. Her professors from Berkeley, including Oppenheimer, Lawrence, and Segrè, were all there at the annual meeting. She introduced these physicists and mentors to her favorite Chinese teacher, Hu Shih, in the hotel where the meeting was held.[14]

Hu was then the Chinese Ambassador to the US. It was near the height of the Sino-Japanese War (but before the Pearl Harbor attack) and he was extremely busy soliciting assistance from the US for the Chinese resistance. In a letter to Adina, Wu told her friend that Hu, just returning to DC from England, would find time to have breakfast with her several times.[15]

In the same trip, Wu visited the colonial site in Williamsburg, Virginia. She wrote a letter to Adina. The letter, dated May 9, 1941, reveals her thinking about life and marriage:

> Although I sat between two young couples in Williamsburg, I did not envy them. Adina, not all marriages are nice. We have to be objective and factual. Life itself is very interesting and colorful.
>
> My love to Adina, Adina Wiens!!! Gee Gee

In addition to revealing her belief in marriage, Wu particularly wrote her friend's name large and with triple exclamation marks. This showed her playful personality, and passion toward her friends.

The young Wu was not only passionate with her girlfriends, but also innocent and lively in private. There were trains connecting Berkeley and downtown San Francisco. Her friend Margaret recalled that she and Wu loved to sit in the front row, watching excitedly the crossing of the bay, and the approaching wonderful scenery.

Margaret also remembered walking back with Wu to the International House from the physics department one night. They passed a lawn, and there was nobody around. They decided to do somersaults, and rolled around like kids. They were grown-ups then, and Wu was probably wearing a *qipao*. Years later they would chuckle when recalling this incident.[16]

A young Wu naturally wanted to look pretty, and would go shopping with friends occasionally. It rained after a shopping trip with Xu Jing-Yi and they fell accidentally. Wu joked: "It is more important to save your pantyhose than your head!" Such were their priorities. [17]

Wu was brought up in an educated family in China. In addition to the many girlfriends, Wu would carefully make friends with several boys. In fact, she had a private reason for consulting her teacher Hu Shih about marriage and choice in her trip to the East Coast. Although she never confirmed it, her many friends said that Hu did give her advice on the eventual choice of love and marriage in that trip.[18]

Wu had many admirers in Berkeley, but she dated only one other young man, Stanley Frankel. For the author, finding out about this other man was quite accidental. Wu did not want to talk about her romance, and Luke was reluctant to discuss his competitor. On October 12, 1989, the weather in Vancouver was cold and damp. When we sat in the Faculty Club of the University of British Columbia, Wu's classmate George Volkoff excitedly reminisced about the student days in Berkeley. He revealed Frankel accidentally.

Volkoff had previously lived in Harbin for three years with his parents, and had special feelings for China. Although he worked on black holes with the great physicist J.R. Oppenheimer, he said that he was just a good student, and never a good physicist, as he had not made any breakthrough in research.

In fact, Volkoff and his advisor at UC Berkeley, Oppenheimer, proposed the possible formation process of a neutron star, after a massive star had exhausted its thermonuclear energy. The *Oppenheimer–Volkoff formula* in a 1939 paper established the foundation of stellar structure within general relativity. Volkoff returned to Canada after graduation from UC Berkeley. He became the President of the Canadian Physical Society, and Science Advisor to the Canadian Government. He brought much vitality to the Canadian scientific society, and became very well-known there.

The retired Volkoff was very animated and talkative. He brought out an old, small, black-and-white photo with three people standing; — one

With boyfriend Stanley Frankel (left) and classmate George Volkoff in Berkeley.
◀

▶

Partying with classmates
in Berkeley.

◀

In the Berkeley era.

was himself, and the other two, hand in hand, were Wu and Frankel. The petite Wu standing between two tall fellows was shorter by more than a head. She and Frankel were beaming, and Volkoff looked more serious.

Volkoff only mentioned Frankel's name, without elaboration. As the saying goes, "A picture is worth a thousand words." This photo revealed that Wu must have had a fairly unusual relationship with Frankel. Given her rather rigid and proper upbringing, she otherwise would not have taken a photo with him hand in hand.

Frankel and Luke were the two "accepted" admirers courting Wu. As she was very proper and cautious, and had a wonderful marriage with Luke, Wu did not have any other romantic involvement.

To prepare for this book, I interviewed the physicist V. Telegdi at CERN, Geneva in May 1990. He was the leader of one of three experimental groups racing to verify the proposal by Lee and Yang of parity violation in 1957. He was also famous for his sharp tongue, inquisitiveness, and practical jokes.

When asked to compare Wu and Madame Curie, the eloquent Telegdi made an interesting assessment: "Madame Curie had a rich love life, but Ms. Wu did not." This serves as a funny footnote of her romantic relationship.

The relationship between Wu and Stanley started in the graduate school at UC Berkeley. Stanley was a brilliant, young, Jewish theoretical physicist working for Oppenheimer. Wu recalled that Stanley was brilliant, the most brilliant student she had ever met. He did not study hard, was sometimes forgetful in examinations, but always got good grades. As for the reason that he did not achieve very much in physics, Wu guessed: "Stanley might be too casual."[19]

Stanley later followed Oppenheimer and joined the Manhattan Project. He worked with the genius physicist R. P. Feynman in developing the atomic bomb at the Los Alamos National Laboratory in New Mexico. Feynman was awarded the Nobel Prize in 1965. He referred to Stanley in his 1985 bestseller, *Surely You're Joking, Mr. Feynman!* In this book, there is a chapter where Feynman recalled the critical problem of calculating the amount of energy released in an atomic explosion. The calculation was too complicated, beyond the ability of his research group. He said: "A clever guy called Stanley Frankel figured out that the calculation could be done by an

IBM computer."[20] Stanley wrote a computer program to perform this massive and complicated calculation with a commercial computer. The critical problem of energy released from an atomic explosion was solved.

In the book Feynman called Stanley merely "*a clever guy*" and spent not quite two pages casually discussing this thorny problem. Anyone aware of Feynman's status and manner would agree that this casual reference was indeed unusual praise from a master. Feynman was well-known in physics for his quick mind, sharp tongue, and immense curiosity. Few people can measure up to his caliber. The "clever guy" Stanley was indeed really clever.

There was romance between Wu and Stanley. They went out alone, and Wu took driving lessons from him, all the usual activities of young lovers.[21] Her dear friend, the artist Sun Duo-Ci, wrote in an article that an American classmate of Wu in Berkeley was deeply in love with her, and that the American stayed single. The last point may have been a mistake: Stanley did get married (but later divorced), and his family visited Wu and Luke once on the East Coast.

The other admirer, Luke Yuan, transferred to Caltech in 1937. The contact between him and Wu became less frequent because of spatial separation. His roommate at Caltech, Zhang Jie-Qian, said that Wu and Luke resumed their intimate relationship probably after 1940. Zhang went to Caltech in 1940 to study aerodynamics, as arranged by the world authority T. von Kármán. He was later elected to Academia Sinica in Taiwan.[22]

From 1940 to the Wu–Luke marriage in 1942, Stanley and Luke probably coexisted in Wu's mind. After 1940, the relationship between Wu and Luke became increasingly steady. Wu had a friend, Xu Jing-Yi, from the International House days in Berkeley. Xu recalled Stanley escorting Wu around. Once, when Wu went to see Luke in Caltech, Stanley was depressed and asked Xu to go out for a drink. Stanley did not say much in the bar, and Xu did not drink. Xu felt uneasy, and was concerned that Wu would not appreciate the idea of them drinking together.[23]

Wu wrote her dear friend Adina in August 1941, and revealed her relationship with Luke. Wu and Adina were planning to vacation at the scenic Lake Tahoe that summer. Wu wrote: "I hope to study all mornings during the vacation, and only join you in late afternoons and evenings. Hope you don't mind."

"Mr. Yuan is dying to see me, but I cannot be in both places. If you don't mind, we could invite him to vacation. He is really very quiet and pleasant." She gave Luke's address at the end of the letter, for Adina to extend the invitation.[24]

There was a woman visitor in the house shared by Luke and Zhang Jie-Qian in the spring of 1942. Luke told Zhang afterward that he and Wu were getting married.[25]

Her girlfriends all supported Wu's choice. Adina met Luke for the first time at Lake Tahoe, and told Wu: "Gee Gee, he is your man." She felt that Luke was stable and dependable, and Wu had made the right choice.[26] Margaret also believed that Luke and Wu were a compatible couple.[27]

Wu Chien-Shiung, and Luke C. L. Yuan were married in Pasadena, California. It was a Sunday, on May 30, 1942, a day before Wu's 30th birthday. The wedding was held in the house of Luke's thesis advisor, Robert Millikan. Millikan had been awarded the Nobel Prize for his measurement of the electron charge, and was then President of Caltech. As it was in the middle of the Second World War and the Pacific War, none of the Chinese family members of Wu and Luke could attend. The wedding was hosted by Millikan, and officiated by a priest and Caltech professor.

The wedding was simple but formal, just as they wished. Mrs. Millikan gave a wedding banquet in her garden, with many of Wu and Luke's classmates and friends in the US attending. The President of the Chinese Students Association in Caltech, Qian Xue-Sen, made an 8 mm movie of their wedding. Qian was later the leader and a major contributor of the guided missile and satellite programs in China.

After the wedding, Wu and Luke spent a week at Laguna Beach, south of Los Angeles. As the house their friend lent them was very large, they dragged two friends along for company — on their honeymoon.

One of the friends was Luke's roommate Zhang Jie-Qian, and the other was Bi De-Xian, Luke's classmate at Yenching University. They were reluctant to go as outsiders on the couple's honeymoon, but gave in at last.

Wu and Luke were frugal. They drove a small car, with the two friends squeezed in the backseat for an hour or so.[28]

Zhang recalled that they had a great time that week, dining out every evening, and the boys swimming in the afternoon. Wu did not swim and

The wedding of Wu and Luke in the garden
of the then President of Caltech, Robert
Millikan.

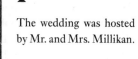

The wedding was hosted
by Mr. and Mrs. Millikan.

was not used to wearing a swimsuit. She was very outgoing, and did not have any of the consciousness or shyness of being a new bride.

After the honeymoon, Luke started working at RCA near Princeton, doing defense work, possibly related to radar. Wu accepted an appointment with Smith College in Massachusetts.

Wu had only teaching responsibility, no research, at Smith College. She missed her friends in California very much, and wrote often to Adina. The letters told of the happy life of a newlywed couple, and their sweet feelings for each other.

In a letter to Adina dated September 19, 1942, Wu wrote: "In these three months living together, I got an even deeper understanding of him. His total devotion and love won my respect and admiration. We are madly in love."

Notes

1. *Wu Chien-Shiung: Queen of Physics*, Sun Duo-Ci, Chinese and Foreign Magazine, 1965.
2. Interview with Zhang Xi-Ying, August 15, 1989, London residence.
   Interview with Yen Mei-He, June 13, 1990, New York City.
3. Interview with Yen Mei-He, June 13, 1990, New York City.
4. Quoted from *Selected Correspondences of Hu Shih*, middle volume, p. 463, China Bookshop, Hong Kong, Chinese Academy of Social Sciences Edition.
5. Same as 4, pp. 512–513.
6. Interview with Wu Chien-Shiung, September 13, 1989, New York residence.
7. Interview with Wu Chien-Shiung, September 13, 1989, New York residence.
8. Interview with Luke Yuan, September 8, 1989, New York residence.
9. Interview with Luke Yuan, September 8, 1989, New York residence.
10. Interview with Wu Chien-Shiung, September 13, 1989, New York residence.
11. Interview with T. D. Lee, who said that he had heard from Mrs. Wilson, October 2, 1989, Columbia University, New York.
12. Interview with Robert Wilson, September 21, 1989, Ithaca, New York residence.
13. Interview with Luke Yuan, September 8, 1989, New York residence.
14. Interview with Wu Chien-Shiung, 1989 and 1990, New York residence. Telephone interview from Taipei with Wu Chien-Shiung, April 5, 1992, New York residence.
15. Letter to Adina from Wu, October 16, 1989. Copied in Adina's residence, Tiburon, California.
16. Interview with Margaret Lewis, February 2, 1990, Library of Harvard Observatory, Boston.

17. Interview with Xu Jing-Yi, October 16, 1989, San Francisco suburb residence.

18. Interview with Xu Jing-Yi, October 16, 1989, San Francisco suburb residence. Interview with Adina, October 16, 1989, Tiburon, California residence.

19. Interview with Wu Chien-Shiung, May 3 and July 31, 1990, New York residence.

20. *Surely You're Joking, Mr. Feynman!*, Richard P. Feynman and Ralph Leighton, p. 125, W. W. Norton & Company, New York, 1985. This book was actually edited and compiled by Leighton (a drummer friend) from a seven-year audio record of Feynman.

21. Interview with Xu Jing-Yi, October 16, 1989, San Francisco suburb residence.

22. Interview with Zhang Jie-Qian, September 26, 1989, Washington, DC suburb residence.

23. Interview with Xu Jing-Yi, October 16, 1989, San Francisco suburb residence.

24. Letter to Adina Wiens from Wu, August 5, 1941.

25. Interview with Zhang Jie-Qian, September 26, 1989, Washington, DC suburb residence.

26. Interview with Adina Wiens, October 16, 1989, Tiburon, California residence.

27. Interview with Margaret Lewis, February 22, 1990, Library of Harvard Observatory, Boston.

28. Interview with Zhang Jie-Qian, September 26, 1989, Washington, DC suburb residence.

Chapter 6

Going East

Wu Chien-Shiung's performance at UC Berkeley was distinguished, but she still had to leave in the second half of 1942 for the East Coast. It wasn't that she wanted to leave, but UC Berkeley could not offer her a teaching position. The reason was simple: there was then not a single female physics professor in the top twenty research universities in the US.[1]

Wu took a teaching position with Smith College in Northampton, Massachusetts. Although Smith was not a research university, it was a prestigious private women's college. A female professor of nuclear physics from the college had visited Berkeley the previous year, and had become a good friend of Wu.[2]

This professor, who was the Dean of Science at Smith College, was eager to recruit good professors like Wu, and she invited Wu to join Smith if the latter ever went to the East Coast. It was wartime, and not many students were in the college. The President of Smith College was pleasantly surprised when eight new students in physics joined the department during Wu's second year.[3]

Wu was quite reluctant to leave Berkeley for Smith, but took the teaching job in Smith College to accommodate Luke Yuan's job situation. Luke transferred to Caltech in 1937. The Sino-Japanese war broke out soon, and he and his Chinese friends were eager to return to China to serve their country. When Hu Shih was in Los Angeles and heard their intentions, he gathered them and gave them serious advice. Hu believed that China would win the war, and would have a great need to rebuild the country. He advised them to stay in the US, learn as much as possible, and be ready to return home after the war.[4] Luke then stayed put in Caltech, obtained a Ph.D. in 1940, did research for another two years, got married to Wu in 1942, and found a position with RCA developing radar for the defense department.

In addition to missing her dear friends in Berkeley, Wu mostly regretted leaving the center of physics research in Berkeley.

Robert Oppenheimer had already assembled a group of top theoretical physicists in the summer of 1942 to start theoretical calculations for the atomic bomb effort in the Manhattan Project. They included the 1967 Nobel laureate Hans Bethe; Robert Serber, Wu's future colleague at Columbia; Edward Teller, the "father of the hydrogen bomb"; Edward Konopinski; and two young physicists — Wu's boyfriend in Berkeley, Stanley, and E. Nelson.[5]

Smith College did not have enough funding to support Wu's beloved research. She was rather depressed, and as a result felt detached from the weather, scenery, and people associated with Northampton.

She wrote to her friends in Berkeley, especially Adina, very often. She missed her friends very much, and was lonely in a new environment.

In a letter to Adina dated September 19, 1942, Wu began:

Dearest Adina,

I missed the nice weather and scenery in California, but I missed you most. Nothing can replace you in my heart. Northampton is not bad, but has many middle-aged women and spoiled, rich girls. I am really lonely here.

In the letter, she complained that Northampton did not have many sunny days, and worried about the coming cold winter, joking that she might not survive. Even the supermarket there irritated her. She complained that it was terrible; a dozen eggs cost 62 cents, tangerines were as costly as museum pieces, and her favorite grapefruits were nowhere to be found.

This letter talked about Luke and her happily married life, and their love for each other. Wu also mentioned news of other friends: Eda had recovered recently from her illness and she asked Adina to visit her on weekends, her former boyfriend, Stanley, had earned his Ph.D. in August and followed Oppenheimer to do defense research, and Hu Shih lived in New York after resigning from his position as US ambassador.

Wu taught in Smith College for a year. She would latter recall that Nancy Reagan (future second wife of US President Ronald Reagan) was there studying drama, and that the female physicist Xie Xi-De (future

President of Fudan University in China) earned a Master's degree there. That year was a rare leisure time for Wu, who was normally obsessed with physics research. She enjoyed other pleasures in life besides science.

Wu wrote Adina in February 1943 congratulating her on her coming wedding, and selected a sterling tea set as a wedding gift. She wrote again on March 30 and thanked Adina for sending California tangerines, reminisced about their happy life in Berkeley, counted blessings of both their happy marriages, and said that she might find a job in New York or Princeton, so that she could be closer to Luke and take better care of him.

Wu said that they were looking for a house in Princeton, and wrote: "Luke was trying to find me a women's bicycle and get me to do more exercise. He wanted to take pictures of me learning to ride a bicycle, and save them for our children to see. I think he is being very naughty."

She continued: "I worry about his health. He told me that he has been working in the laboratory day and night, including Sundays, for the last several weeks. He told me not to worry. He has a good appetite, and plays badminton once a week in a local club. I think he will beat everybody pretty soon."

Wu wrote at the end: "Did you hear the speech of Soong May-Ling (the first lady of China) on the radio? She was really eloquent. We Chinese are proud of her."

She asked Adina to visit Eda, who was working as a chemist in the Radiation Laboratory. She also said that she had not written Eda since leaving Berkeley, and must do so that evening.

In this period, Wu's only way of keeping up with the progress in physics was through reading physics periodicals and attending conferences. She met Lawrence at a conference in Boston. He was concerned about her situation, and she told him frankly of her frustration of not being able to conduct research. Lawrence, with his Nobel Prize, was influential in physics. He immediately wrote letters of recommendation to several universities.[6]

Eight universities, including Princeton, Brown, Harvard, MIT, and Columbia, all wrote to offer Wu a position. In 1943, US universities were discriminatory toward hiring Jews, Asians, blacks, and women as professors, and some did not even accept women students. Even though Smith College promoted Wu to associate professor, and gave her a huge pay raise in the second year,[7] she decided to go to Princeton, and became the first female instructor in history there.[8]

There was a side story in her going to Princeton.

When in Harvard accepting the *Lifetime Achievement Award* in 1983, she recalled this story in her lecture: she was very happy accepting the appointment as instructor in physics from Professor Henry Dewolf Smyth of Princeton in 1943; she did not hear anything for several weeks, and finally received a letter from Smyth, saying that he had not realized how difficult it was to employ a woman in teaching.

Wu and Luke had a home at 306 Nassau Street, Princeton. Luke also moved to Princeton University, as a research associate, in 1946. They had a happy time there and their only son was born in Princeton.

Wu taught physics to naval officers at Princeton University. There had been rapid progress in physics and the officers were scared of new courses such as quantum mechanics. Among the students was J. Fletcher, who later served as the Director of NASA.

Wu still had no opportunity to do experiments. Several months later, Columbia University started a Division of War Research and Lawrence recommended Wu for an interview there. The interview was conducted in the Jet Propulsion Department. Two physicists quizzed Wu on many problems in physics for a whole day, but were careful not to reveal their secret plans.

At the end, they asked: "Now, Miss Wu, do you know anything about what we're doing here?" "I'm sorry," she said smilingly, "but if you wanted me not to know what you're doing, you should have cleaned the blackboards." They burst out laughing and suggested: "Since you already know what's going on, can you start tomorrow morning?"[9]

Wu and Luke started living in Princeton in 1943, and met many Chinese there. Quite a few of them became lifelong friends. It was near the height of the Sino-Japanese War; these Chinese faced an uncertain future, and were unable to return home. Yet, as youths, they had boundless confidence and optimism.

Among the Chinese friends, there was the internationally famous architect I. M. Pei, the world-authority mathematician Chern Shiing-Shen, first-rate economist Fang Shan-Gui, and Zhang Xue-Ceng, the brother of Zhang Xue-Liang. At the Institute for Advanced Study, there was the top Chinese mathematician Hua Luo-Geng, the physicists Zhang

Wen-Yu and Hu Ning, and the overseas academician Zhang Jie-Qian. In addition, the Chinese physicist Yu Nao-Tai and Shu Guo-Hua (who later served as the Premier of Taiwan) stayed in Princeton for a short time, and socialized with Wu and Luke.[10]

I. M. Pei graduated from the architecture school at MIT, and also worked on national defense in Princeton. His research project was to find the most effective way to bomb and destroy houses.[11]

In November 1989 Pei reminisced about the events forty years before. He sat behind a desk in his modest office on the ninth floor of a building on Madison Avenue in Manhattan, wearing his signature round black eyeglasses, and a constant warm smile. He said that the US had declared war on Japan after the Pearl Harbor attack, and China and the US were allies. The Chinese should join the war effort, and the best way was to work in a national defense project.

He was married for a year in 1943, and lived in a rented house on a chicken farm. He painted and decorated the house in a special way, which put his talent on display.[12] He said that they had no money then, and just had to do something different.

Pei was born in Guangzhou, when his father Pei Zu-Yi was sent there by the Bank of China, but he grew up in the Suzhou and Shanghai area. He and Wu could converse in the Shanghai dialect. Pei said that he and Wu had hit it off very well from the beginning. Wu was easy-going, and without any arrogance associated with a successful scientist. She was factual and concise, but had a humorous side too.[13]

These friends gathered in the residence of Wu and Luke on weekends, had a miniconference, and played bridge after dinner. Pei and Chern would be there. Wu did not play bridge, but might cook a couple of dishes. Luke was a fairly good cook. He singled out dumplings, chicken, sautéed vegetables and "lion's head" (a pork meatball dish) as her favorites.[14]

Wu later recounted these happy days in a speech in English delivered at the sixth annual meeting of the Eastern American Association of Chinese Scholars in 1981.

She said: "Today's big meeting is a far cry from our mini-conferences of ten or so people in Princeton in those days. We frequently met in our house for informal discussions on Saturday mornings, and then drove to the only French restaurant in town for lunch. As the only means of travel was our old Buick, we would all squeeze in with some difficulty. We did all right once inside as all Chinese were skinny, but were quite embarrassed about getting off one by one, like a string of clowns in a circus."

She added: "To this day, we still remember those happy days and funny events in Princeton."[15]

Apart from the Chinese friends, there were many foreign scientists as guests in the Wu residence, particularly Pauli and von Kármán.

Pauli was honored as "The Great Pauli", a first-rate scientist who was also famous for his blunt remarks. He once told the "father of the hydrogen bomb", Edward Teller, to his face that Teller was no longer a scientist, but more a politician. Teller was not pleased.[16] Pauli and Wu hit it off extremely well because they had so many common interests in physics. Pauli was very considerate, and asked people not to discuss the topic of war with Wu, because he knew about the war experience of her family and did not want to hurt her feelings.[17]

von Kármán was a world authority on aerodynamics. He had two famous Chinese students. One was Qian Xue-Sen, who was instrumental in the guided missile and satellite program in China. The other was Lin Jia-Qiao, a member of Academia Sinica in Taiwan. The "guest" on Wu's honeymoon, Zhang Jie-Qian, was his student also. von Kármán was always very nice to Luke. He left his house in Pasadena and his car for Luke's use when he took a vacation in France every year. Luke recalled that von Kármán loved smoking cigars but was forgetful. He would light one, put it away while talking to a visitor, and light a new one when he started talking to another visitor.

Pauli and von Kármán both studied physics in Germany, and they knew each other well. von Kármán switched to applied aerodynamics. He had a girlfriend during the student days, who got married to a theorist. von Kármán joked that she preferred theorists to applied scientists. He remained single all his life.[18]

When Pauli and von Kármán went to dinner at Wu's residence, von Kármán, who loved French wine, brought a bottle, and left it outside in the snow to chill. They chatted after dinner and forgot about the wire. When Mrs. Pauli was ready to leave, Pauli remembered the wine and teased von Kármán: "Is it cold enough now?"[19]

Wu and Luke's married life was very sweet. Wu wrote Adina on October 2, 1943, saying that they had a fireplace in the apartment. Luke bought a bunch of firewood and joked that they probably were the descendants of Eskimos. Wu said that it was so peaceful and relaxing sitting by the fireplace.

She also said that Luke planted Chinese cabbage, corn, tomato and watermelon in the garden at work. She cooked when a large group of old friends visited them in Princeton, and luckily no one complained about her cooking.

Wu worried about the safety of her immediate family remaining in China in the heat of the Sino-Japanese War. Her mother and maternal grandmother had stayed in Shanghai, which was now occupied by Japan, and her father and uncle had retreated to Chongqing to participate in the construction of the Dian Mian Road (Burma Road).

She also mentioned a parcel of medicine that she had attempted to send to her father in Chongqing. As there was no postal service between the US and China, the parcel was mailed from Princeton, and went as far as San Francisco before being returned.

As Adina had a brother, Harold, who was in Chongqing with the US Navy, Wu asked for a favor whether Adina could send the parcel to Harold as a Christmas gift, then have it delivered to her father. Adina made her wish come true.

Wu started working in wartime research at Columbia University in March 1944. She lived in a dormitory near the medical school at West 168th Street, and returned home in Princeton for the weekends. She was thrilled to return to physics research, and hoped that the effort would make positive contributions to the war and China. Wu and Luke still planned to return to China soon.

Japan surrendered in August 1945, and Wu finally received a letter from her family in China. She wrote Adina on December 13, 1945: "We have not firmed up our plan to return to China. It is just heartbreaking to think of the stupidity going on in our country." Although Chiang Kai-Shek (the President) and Mao Ze-Dong (the revolutionary) had announced the agreement in their joint meeting on October 10, the Kuomintang Government and the Communist Party were on the brink of an all-out civil war. The possible civil war was the "stupidity" she referred to in her letter.

Wu also mentioned a puppy she sent to Adina's daughter Charlotte, and the safe delivery of her newborn son, Daniel. She was happy to stay home on weekends — now Friday night to Monday after the war, instead of only Sunday like before.

She also wrote that her friend Stanley from Berkeley had visited them in Princeton: "Stanley and his wife, Mary, had a business trip to the East, and they spent a Sunday with us in Princeton. When I sent them off in the train station, I could not hold back my tears. It was nice to have a reunion with old friends. Dear Adina, when shall we see each other again?"

Wu got pregnant. Her colleague J. Winkle recalled that she had terrible morning sickness.

She gave birth to a boy in Princeton Hospital on February 15, 1947. The baby was very late, and the labor lasted for a whole day. Wu had a cesarean section and lost much blood. They named him Yuan Wei-Cheng, with the English name Vincent.[20]

Wu stayed in the hospital for a while after giving birth. Albert Einstein visited her once, as his sister was also being treated in the hospital. Luke was not around, and Einstein had a heavy German accent. His description of a "cesarean section" was "cutting open." Rather horrifying.[21]

The childbirth was a severe blow to Wu's health. She assumed Vincent's voice when writing to Aunt Adina: "Mother stayed in the hospital for three weeks, lost a lot of blood, had two blood transfusions. She was very weak but is now recovering well."

She added: "I was born 7 lb. 14 oz., length 21 in., and I am two months old. I might not care for the sour orange juice, but I must have made a funny face when forced to drink it, as my mother laughed so loudly. She discussed it with the pediatrician, and changed the orange juice to a solution with vitamin C."[22]

Her work did not slow down with the arrival of the newborn. She earnestly continued her research in nuclear physics at Columbia, with Vincent taken care of by a competent nanny. The newborn brought much joy, and her letters bared all her motherly love.

In a letter to Adina dated December 14, 1947, she wrote a whole paragraph about Vincent: "Vincent is a lovely little monkey. He is funny." She said that Vincent loved to play with Luke and her, and described their trick of placing an object far away and challenging him to pick it up with

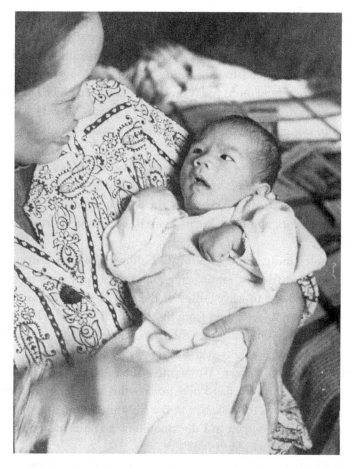

A happy mother at the birth of son Vincent.

chopsticks. She wrote: "He picked it up and immediately showed us the winning smile. He is a riot. I am sure you will like him. He is still shy. He will be ten months old tomorrow, has eight teeth, can stand and sit up. He calls himself Di Di (younger brother)."

Wu's was getting busy with her research work. She had time to write just before Christmas. She wrote Adina on December 16, 1948: "Vincent is almost two. He is naughty, and loves to tease others. But he is the most rational child I know of. When I come home for weekends, he follows me around the house and is not even willing to play outside. He is called 'Mommy's boy.'"

Luke left Princeton in 1949, and joined the Brookhaven National Laboratory in Long Island, New York. They bought a house in Long Island, and Wu also got a university apartment from Columbia University.

Vincent started going to a boarding school on Long Island in 1953. Luke sent him to school on Monday mornings, and then picked him up and spent weekends in the Columbia apartment. Vincent did very well in his first year and skipped two consecutive grades. He was not quite seven years old, but was in third grade already.[23]

Vincent transferred to the Collegiate School in Manhattan the next year, and lived with Wu. He took the school bus at 8 a.m. and returned home at 4:30 p.m. Wu always stayed in the laboratory after 6 p.m. Vincent would do homework and take care of himself. He was in fourth grade, and "very lovely".[24]

The whole family had a vacation in California in the spring of 1955. Wu met many old friends, reminiscing about the happy days in the past. Vincent did not want to leave Adina at the end of the vacation. Wu met Stanley, who was now separated from his wife, Mary, but was happy working in Los Angeles. They also met their "matchmaker", Victor Yang, who was married and had an eight-year-old son.[25]

At the end of that year, Wu wrote Adina about her busy life and Vincent's school: "Vincent has a new teacher this year. This teacher complained that Vincent is too talkative in the class and frequently plays tricks to get attention. Well, what can I do? Should he be punished because he is happy, energetic, and naughty? No!"[26]

Wu and Luke had been planning to return to China in the previous years, but the civil war in China was getting worse every month. The Communist Party had total victory in 1949, and controlled the entire mainland. Chiang Kai-Shek (the President) and the Kuomintang Government retreated to Taiwan. Wu's father urged them not to return, and they also did not want Vincent to grow up in a communist country. They decided to stay in the US.

Wu's research achievements in Columbia became well known throughout the world, and she was promoted to associate professor in 1952. She was very busy even without doing much housework. There was a Chinese friend Lu Gwei-Djen from the Berkeley days. Lu was then working in the Columbia medical school. She later moved to the University of Cambridge in England, working with Joseph Needham, a British historian

1948 in the US.

who wrote *Science and Civilisation in China*. They were married after Needham's wife died. Lu died in 1991, and Needham in 1995.

Wu recalled that she and Lu would buy fabric and sew their own *qipao*. During the Second World War, and then the civil war in China, they could not get a supply of new *qipao*. Wu wore the dress all those years, a token of keeping her memory of the mother country.

She had to attend conferences in other countries, but she and Luke were holding passports issued by the Kuomintang Government in Taiwan. In 1952, Wu was invited to deliver a plenary report on beta decay at the International Congress in Holland. She could not get a visa, as a Republic of China passport was not recognized. Only through the assistance of Pauli from Switzerland was she able to make the trip.[27]

All these inconveniences made Wu change her mind and she became a naturalized US citizen. The Chairman of the Columbia physics department, 1964 Nobel laureate Charles Townes, obtained a special citizenship quota from the US Congress for Wu.[28]

Wu and two of her students who served as witnesses went to the Immigration and Naturalization Office near Chinatown in 1954. The officer first had to verify the qualification of the witnesses, and asked them if they had any criminal record. One honestly reported getting a traffic ticket before, and was told that it did not count.

The officer proceeded to ask Wu simple questions, such as: "What is democracy?" Wu was all serious: "I am sorry, but I have to sit down in order to answer your question in detail." The officer laughed: "There is no need to do so. Just a short answer would be fine."[29]

Wu had reluctantly left California to work in the East, and ended up staying at Columbia University permanently. She had planned to return to China, but the political changes there made her stay in the US and she became a naturalized citizen. Her talent and persistence propelled her to the pinnacle of physics research in the world, but life's happenings would keep her in the US.

In the thirteen years after she left California, Wu had a family, a son, and entered a new stage of life. In science, her achievements were of world-class caliber. And better things were yet to come.

Notes

1. *Nobel Prize Women in Physics*, Sharon B. McGrayne, Carol Publishing Group, New York, 1993.
2. Interview with Wu Chien-Shiung, October 26, 1990, New York residence.
3. Interview with Wu Chien-Shiung, July 25, 1990, New York residence.
4. Interview with Luke Yuan, September 8 and June 19, 1988, New York residence.
5. *A Mind Is Always in Motion: The Autobiography of Emilio Segrè*, Emilio Segrè, UC Berkeley Press, 1993.
6. Interview with Wu Chien-Shiung, July 25 and August 28, 1990, New York residence.
7. Letter to Adina from Wu, October 2, 1943, provided by Adina.
8. Interview with Wu Chien-Shiung, July 25, 1990, New York residence.
9. *Nobel Prize Women in Physics*, Sharon B. McGrayne, Carol Publishing Group, New York, 1993.
10. Interview with Luke Yuan, September 10, 1989, New York residence.
11. Interview with Wu Chien-Shiung, September 13, 1989, New York residence.
12. Interview with Wu Chien-Shiung, September 13, 1989, New York residence.
13. Interview with I. M. Pei, November 29, 1989, Office of Pei Architect, New York.

14. Interview with I. M. Pei, November 29, 1989, Office of Pei Architect, New York. Interview with Luke Yuan, September 10, 1989, New York residence.
15. Speech of Wu Chien-Shiung in English; the translation was published in *Reading Times*, second edition, December 10, 1981. The quotation was from the translation.
16. Interview with Luke Yuan, May 5, 1990, New York residence.
17. Interview with Wu Chien-Shiung, January 27, 1990, New York residence.
18. Interview with Wu Chien-Shiung, May 5, 1990, New York residence.
19. Interview with Wu Chien-Shiung, February 12, 1990, New York residence. Interview with Luke Yuan, May 5, 1990, New York residence.
20. Interview with Luke Yuan, September 10, 1989, New York residence.
21. Interview with Luke Yuan, September 10, 1989, New York residence.
22. Letter to Adina from Wu, April 18, 1947, provided by Adina.
23. Letter to Adina from Wu, November 5, 1953, provided by Adina.
24. Letter to Adina from Wu, November 14, 1954, provided by Adina.
25. Letter to Adina from Wu, April 30, 1955, provided by Adina.
26. Letter to Adina from Wu, December 1, 1955, provided by Adina.
27. Interview with Wu Chien-Shiung, September 20, 1989, New York residence.
28. Interview with Charles Townes, March 29, 1990, office in UC Berkeley.
29. Interview with Wu Chien-Shiung, December 14, 1989, New York residence. Interview with Noemie Koller, December 15, 1989, office in Rutgers University, New Jersey.

Chapter 7
From Nuclear Fission to the Manhattan Project

Very few scientific discoveries have had such an immediate impact on society like nuclear fission. Common discoveries do not leave lasting impression. The reason that the discovery of nuclear fission shook the world was the clear demonstration of the destructive power released by the atomic bomb. This kind of demonstration was unfortunate, but very effective.

Development of the atomic bomb began with the start of the Manhattan Project in the US. This historic project recruited all the top scientific minds of the Allies in the Second World War. Among them was the distinguished Chinese physicist Wu Chien-Shiung.

Her participation was remarkable not only because she was foreign and female, but also because her role was a critical piece of the project. Wu had made important achievements in nuclear research, and the head of the project — the "father of the atomic bomb", Robert Oppenheimer — appreciated and valued her as a former student and for her work. Though a noncitizen and a fairly new arrival in the US, Wu was nevertheless able to get special permission and top-secret clearance to join this defense work.

The story of Wu and the Manhattan Project starts with her first arrival at UC Berkeley from China in 1936.

Nuclear physics research was in full bloom. Several discoveries in the previous years had propelled nuclear physics into the most challenging frontier.

The British scientist J. Chadwick discovered neutrons in nuclei in 1932, and confirmed the atomic model consisting of protons and neutrons inside the nucleus and electrons sourrounding the nucleus. The French scientists Irene Curie (daughter of Madame Curie) and her husband,

Frederic Joliot, were doing similar research in Paris. They missed the discovery of the neutron, but won the Nobel Prize for the discovery of artificial radioactivity in 1934, as predicted by the British physicist E. Rutherford, who had said: "Joliot and Irene are so brilliant, they will soon win a Nobel with a different achievement."[1]

At the same time in Rome, the great physicist Enrico Fermi also worked on using neutrons to produce excited states in many elements. He had made important contributions to both theoretical and experimental physics. His pioneering works broadened the field of nuclear physics research, and established the foundation for the later discovery of nuclear fission, and the era of atomic energy.

Before her arrival to San Francisco from Shanghai by boat, Wu was a teaching assistant at Zhejiang University, and performed experiments at the Academia Sinica at Nanjing. She was exposed to the aforementioned exciting discoveries, even though she was not directly involved.[2]

Wu was planning to stay in Berkeley for a week, and continue to the University of Michigan, but she became attracted by the vitality and energy at UC Berkeley.

The vitality at Berkeley hinged on the collection of brilliant young scientists, like Lawrence and Oppenheimer. Their scientific talents made Berkeley a new center of research, breaking the monopoly of the East Coast. The cyclotron invented by Lawrence became the best research tool at that time.

When Wu visited the physics department with Luke Yuan as a guide, she recognized the boundless potential there. She was confident about the future, had great expectations for herself, and decided to stay in this challenging environment. Wu, accompanied by Luke, went to see the Chairman of the physics department, R. Birge. In the interview, Birge recognized Wu's unusual talent in physics, and made an exception to accept Wu into the graduate school even though the academic year had already started.[3]

After two years of course work, Wu was ready to begin her Ph.D. thesis work. The Director of the Radiation Laboratory, Ernest Lawrence, became her official advisor in 1938.

The Radiation Laboratory was built around the cyclotron invented by Lawrence. He first succeeded in 1931 in constructing a circular accelerator with a magnet having a diameter of a couple of inches. He gradually scaled up the accelerator, with the diameter of the magnet increased to 9″, 11″, and 27.5″. By 1936, when Wu started in Berkeley, the cyclotron's magnet was 37 inches in diameter. The energy of the accelerated charged particle increases with the size of the magnet. The bombardment of target nuclei with these high-energy particles would produce more fragments to be studied. The cyclotron had become the best tool in nuclear research at that time.

Wu used the 37″ cyclotron in her experiments. It accelerated the positively charged protons to high energy, bombarded different targets with these protons, and produced neutrons for further probing of the nuclei. It also accelerated deuterons (which consist of one proton and one neutron).

The 37″ cyclotron could produce deuterons with energies of 8 million electron volts (MeV), and the 60″ cyclotron built in 1939 produced deuterons with an energy of 16 MeV. Many unstable, radioactive isotopes were produced in the bombardments by these high-energy particles. Many artificial elements were discovered afterward.

These discoveries elevated Berkeley to a world-renowned research center for nuclear physics within 10 years, and Lawrence won the 1939 Nobel Prize in Physics for his invention of the cyclotron.

From the very beginning, Wu distinguished herself by her experimental work. Between 1938 and 1940, she completed two nice projects. Lawrence directed her first experiment that investigated two modes of X-rays excited by the electrons emitted in the beta decay of radioactive lead. Wu produced unambiguous experimental evidence and analyses that would differentiate among various theories. She succeeded in understanding the disagreement between theoretical calculations and similar experimental results obtained by other scientists.[4] From the very first experiment, she demonstrated this research style and quality of work — relentless pursuit of precision — which was sustained throughout her life to come.

Her accuracy and persistence in doing experiments, and her insight into physics, were highly praised by others. L. W. Alvarez (1968 Nobel Laureate in Physics) was a postdoc in Berkeley who arrived there slightly before Wu. He wrote in his autobiography:

> It took many hours waiting for the sample to decay in a radioactivity experiment. I got to know this graduate student in this idle time. She used the same room next door, and was called "Gee Gee." She was the most talented and most beautiful experimental physicist I have ever met.[5]

Wu's second project was related to nuclear fission, and later became greatly relevant to the production of the atomic bomb.

Lawrence was her official thesis advisor. He managed the Radiation Laboratory, and spent much of his energy and time to solicit funding for the expansion and the operation of the cyclotron. Although Wu frequently met with him and got his advice on her experiments, her actual advisor was the Italian physicist Emilio Segrè.

Segrè worked with Fermi in neutron beam experiments in Rome in 1934, and was a seasoned experimentalist. Without a teaching position in Berkeley, he could not officially take on thesis students. But, in fact, Wu was his first Ph.D. student in Berkeley. Segrè would later win the Nobel Prize in Physics in 1959, with the discovery of the antiproton.

Wu and Segrè started in 1939 to investigate the fission products of uranium. In particular, they identified the inert gas xenon as one of the products. Xenon had a critical effect on the chain reactions in uranium fission.

The investigation by Wu and Segrè began with the discovery of nuclear fission by two German scientists in 1938.

In the history of scientific discoveries, many, including nuclear fission, were quite accidental. From 1934 to 1938, there were three groups in Europe as the world center of research at that time that were studying the possibility of splitting nuclei. Fermi led the group in Rome; the daughter and son-in-law of Madame Curie led the group in Paris; and Otto Hahn and Lise Meitner led the group in Berlin.

After Hitler occupied Austria in 1938, Meitner, who was an Austrian Jew, was exiled to Sweden through Holland and Denmark. Hahn was left with a German chemist, Fritz Strassmann. Before publication, Hahn and Strassmann would send to Meitner in Sweden any new result for review.

When her nephew, the physicist Otto Frisch, visited Sweden in the Christmas of 1938, Meitner showed him the recent results of Hahn and Strassmann. They discussed the results and believed that the observation had shown a uranium nucleus splitting into two — nuclear fission. It was a shocking discovery.

Frisch returned to Copenhagen and met the great Danish physicist Bohr, who was ready to depart for a visit to the US. Bohr realized the obvious as soon as Frisch started reporting the new result of nuclear fission.[6]

Bohr knew that this discovery was very significant, and tried to keep it secret and known only to a small circle of people. Not long after his arrival in the US, the news of nuclear fission started to circulate in the scientific community. Meitner and Frisch published their paper discussing nuclear fission in the January 16, 1939 issue of *Nature*. Nuclear fission officially became an earth-shaking open secret.

Once uranium fission was discovered, every scientist in this field rushed to start relevant experiments. Wu and Segrè investigated uranium fission products using the neutrons produced in the cyclotron at Berkeley. The experiments ran from 1939 to 1941 and had many important results. At the beginning Segrè provided many ideas, but Wu completed much of the later work.

They used the 37″ cyclotron to accelerate deuterons to 8 MeV, and later the new 60″ cyclotron to accelerate deuterons to 16 MeV. Then, by colliding the accelerated deuterons with a beryllium target, they produced neutrons. The resultant neutrons collided with uranium or thorium nuclei to induce nuclear fission. They then investigated the fission products.

During those years, Segrè once took leave from Berkeley and visited Fermi and other scientists in New York. He told Wu that he went fishing for "inspiration".

Wu worked alone when Segrè was away. She identified two radioactive (inert gas) xenons in the iodine produced in uranium fission, and measured their half-lives, radioactivity, and isotope mass numbers. When Segrè returned from the East Coast, he was very pleased with the results, and believed that Wu was capable of doing first-rate independent experiments.

She wrote up the results listing Segrè and her name for publication. Segrè read the paper but deleted his name, as he thought that Wu should take all the credit. The paper was published in the *Physical Review* with Wu as the sole author.[7] It was 1940.

Wu earned a Ph.D. degree in 1940 with two distinguished experiments to her credit. A usual Ph.D. research may not be that original. Wu did not just one but two creative experiments.

Her thesis defense was held in the Berkeley physics department's Le Conte Hall on Friday, 17 May 1940. Robert Wilson, who accompanied her to the meeting, recalled that Wu was so nervous that she almost fainted. Seaborg (1951 Nobel laureate in Chemistry) entered in his diary:

> The thesis defense of Wu Chien-Shiung was held in room 222, Le Conte
> Hall at 4 pm. Her committee consisted of Lawrence (Chairman), Robert

B. Brode, Seaborg, Alvarez and Leonard B. Loeb. All members agreed that her performance was very good.[8]

Her committee included Lawrence, who had already won a Nobel Prize, and Seaborg and Alvarez, who would both win the prize later. It was quite a gathering. Wu was highly praised by the committee members.

Wu's classmates, such as Glenn Seaborg and Robert Wilson, recalled her ambition and determination. Wilson said: "Chien-Shiung was very ambitious. She frequently quoted Madame Curie as her role model. She wanted to excel. I could feel her determination, and was confident that she could accomplish whatever she wanted."[9]

Wu recalled that the cyclotron was also used for radiological treatment of cancer patients. She had to schedule her experiments properly, so as to avoid conflicts. The machine operators would specially arrange the radiation beam for her experiments.[10]

Wu's talent was highly praised by her professors. Her mentor and collaborator Segrè maintained a close friendship between student and teacher. He wrote in his autobiography: "Wu's will power and devotion to work are reminiscent of Marie Curie, but she is more worldly, elegant, and witty."[11] Her advisor, Lawrence, believed that Wu was the most talented female experimental physicist he had ever known, and that she would make any laboratory shine.[12]

In early 1939, several great physicists in the US had different opinions on whether the knowledge of nuclear fission should be freely circulated, just like other scientific knowledge. The Hungarian scientist L. Szilard insisted that it should be kept secret. Fermi felt otherwise at first, as his own calculation showed that chain reaction was impossible since there were not enough neutrons produced in nuclear fission to trigger further reactions.

A letter of Joliot and Irene Curie published in *Nature* on April 22 settled these arguments. They confirmed that each nuclear fission produced three-and-a-half neutrons on the average. Chain reaction was feasible.

The Second World War was getting intense, especially in the European theater. Germany, seeking every advantage in the battlefield, showed keen interest in the nuclear fission first observed in Berlin. The researchers in the US would subconsciously keep the most sensitive results of nuclear fission secret. Segrè and Wu studied the fission products of uranium, and

Wu doing an experiment in the 1940s.

published the results in two papers in 1940. They kept the knowledge relevant to chain reactions secret, and intended to publish it in another paper only after the war in 1945.

Oppenheimer was the undisputed leader of the theory group in the Berkeley physics department, and he attracted a group of the most brilliant minds. These scientists were the main force behind the successful development of the US atomic bomb, and the postwar rise of the US sciences.

This group met every Monday evening in the basement library of Le Conte Hall. Oppenheimer would discuss his new theories, or other young scientists would present their new experimental results.

One day, the Berkeley physicists wanted to hear about the latest development in nuclear fission, and Oppenheimer invited Wu to speak. She first discussed the pure physics of nuclear fission for about an hour, and then mentioned the feasibility of a chain reaction. She said: "I have to stop here, can't go further."[13]

Lawrence in the audience, burst into laughter, turned around and saw Oppenheimer also smiling. They both knew what Wu meant. Although she stopped the lecture, she continued with her research.[14]

That was the dawning stage of nuclear fission. Wu gained understanding from her own experiments, and reviewed and compiled the latest developments. Her lectures were very insightful. Whenever Segrè was invited to lecture on nuclear fission, he would ask to borrow her lecture notes. Oppenheimer also appreciated her deep insight into this field, and always said: "Go invite Ms. Wu — she knows everything about the absorption cross section of neutrons."[15]

Based on her achievements in Berkeley, Wu was honored as the "Chinese Madame Curie" in China. The *Oakland Tribune* in Berkeley published an article on April 26, 1941 entitled "Outstanding Research in Nuclear Bombardments by a Petite Chinese Lady". The article included a rather large picture taken from her passport. Wu in the picture had bright eyes, a delicate mouth, and a confident expression on her pretty face. She was a charming lady.

This news report said: "A petite Chinese girl worked side by side with some top US scientists in the laboratory studying nuclear collisions. This girl is the new member of the Berkeley physics research team. Ms. Wu, or more appropriately Dr. Wu, looks as though she might be an actress or an artist or a daughter of wealth in search of Occidental culture. She could be quiet and shy in front of strangers, but very confident and alert in front of physicists and graduate students."

It said that when immersed in a nuclear fission lecture in front of physicists, Wu wrote a formula backward (right to left, like Chinese), and truly impressed the audience.

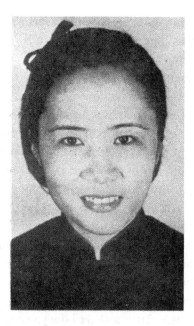

Photo of Wu Chien-Shiung published in the *Oakland Tribune*.

After a brief discussion on her upbringing, her education in China, and her study in the US, the news article ended:

China is always on her mind. She was so passionate and excited whenever "China" and "democracy" were referred to. She is preparing to return and contribute to the rebuilding of China.

The attack on Pearl Harbor started the Pacific War. This and other factors changed her plan.

The scientists of the Allies discussed the feasibility of an atomic bomb, but never started the development. The Allies suffered many defeats on the battlefields, and then Germany imposed an embargo on any shipment of uranium from the occupied Czech Republic. The Allies suspected that Germany might have started serious atomic bomb development.

A German scientist, Siegfried Flugge, surprisingly published a paper in a German periodical making public the recent results in nuclear fission. He was just trying to beat the censorship, but the scientists of the Allies misread the signal, thinking incorrectly that Germany must be much further along if they were willing to publish the results. This event accelerated the US project to develop an atomic bomb.[16]

The Hungarian scientist L. Szilard decided to take action, and he suggested that the first step was to take control of the uranium mine in Congo (a colony of Belgium). As Einstein was an acquaintance of the royal family of Belgium, his help was solicited. Szilard and the banker A. Sachs then sent a letter to President Roosevelt, urging him to develop an atomic bomb in the US. To add credibility to the letter, they asked Einstein to co-sign. This letter was critical in the development of the atomic bomb, and left Einstein with much regret after the war.

The "Manhattan Engineering District Project" started in June 1942; it was so called as its headquarters was located in Manhattan, New York. The first Director was General G. Marshall, who was later replaced by General L. R. Groves. The Scientific Director was Oppenheimer.

Wu and Luke Yuan met each other in 1936 in Berkeley, and got married in Pasadena in May 1942. After a short honeymoon, Luke went to RCA near Princeton, doing defense research on radar, and Wu went to Smith College near Boston as an assistant professor.

Wu stayed at Smith College for one year, and moved to Princeton University as an instructor, teaching Physics to naval officers in 1943. Starting March 1944, she went to Columbia University as a Senior Scientist. She received top-secret clearance as a non-citizen, and participated in the Manhattan Project.

The US atomic bomb project was pretty far along in 1944. Scientists started to grasp the tremendous power released by such a bomb. There were two critical outstanding problems: one was a way to concentrate uranium to achieve a critical mass; the other was a technique to effectively initiate the chain reaction. Wu's work in Columbia was on the processes of concentrating uranium, and was also focused on developing a very sensitive gamma detector.[17]

The center of the Manhattan Project was in the Los Alamos National Laboratory in Santa Fe, New Mexico. General Groves and Oppenheimer were there in charge. The main tasks in Los Alamos were the core calculations for the bomb, and simulation and production of the explosion trigger. They had a cyclotron from Princeton disassembled, moved and then reassembled in New Mexico. There were several laboratories where uranium 235 (capable of undergoing fission) was extracted from uranium 238, then further concentrated. The extraction/concentration process was developed in two places: the Westinghouse Laboratory near Pittsburgh, using centrifugation; and Columbia University, using gas diffusion. Both methods were based on the differences in atomic masses and molecular sizes of these isotopes.

There was a new metallurgy laboratory where a nuclear reactor was being built. Several top scientists — the Nobel laureate E. Fermi and the Hungarians Leo Szilard and Eugene Wigner — were in charge of a group of young physicists there. The reactor was used to study the controlled chain reactions in nuclear fission for possible peacetime applications, and to produce a new element (plutonium) capable of undergoing fission.

Putting these development processes together, the real production was carried out in the Oak Ridge National Laboratory in Tennessee, where the extraction and concentration of uranium was performed, and in the Hanford Reactor Factory in Washington State, where plutonium was produced.

Wu took part in the production process of uranium using gas diffusion. The project was held in a converted Nash automobile warehouse in West 136th Street near Columbia University. Another laboratory was in the Pupin Laboratories. The project, codenamed "Special Allies Material", was under the Director John Dunning. Wu was under William Havens, with James Rainwater as a collaborator. The latter would later win a Nobel Prize in Physics. The production process was directed by Harold Urey (who discovered helium), G. M. Murphy, and others.[18]

The military representative in charge of the management issues of the project was Colonel K. D. Nichols.

Fermi started the operation of the Hanford Reactor in Washington State on September 27, 1944. The chain reaction proceeded very well, stopped after several hours, but resumed after another few hours. Fermi and J. Wheeler suspected that this stop-and-go phenomenon was caused by the absorption of neutrons (the trigger for the chain reaction) by some fission product.

They later found out that the suspect was indeed a fission product — the radioactive, inert gas xenon (Xe-135). The xenon isotope has a half-life of 9.4 hours, and a large absorption cross-section for neutrons, causing the stoppage of chain reactions. As a stoppage of chain reactions was called "dead," they termed xenon the "poison gas."[19]

Wu's professor Segrè told them to call her, as he remembered that Wu had a deep understanding of the absorption cross-sections of neutrons by xenon. Fermi and Wheeler sent a telegram to New York.

Colonel Nichols got the telegram and found Wu: "Ms. Wu, I got a telegram from Fermi and Segrè. They would like to have the draft paper on the experimental results you wrote in Berkeley. May I have it?"

Wu originally agreed with Segrè that a paper on the experiment would be published in the *Physical Review*, but the critical and sensitive data would only be released after the war. She said: "I cannot give you the data unless I receive the request directly from Fermi and Segrè."

Colonel Nichols found the head of the theory group, Murphy, who in turn found William Havens whom Wu knew fairly well. Havens said: "We know each other well. Dr. Nichols has been very helpful to us, and Los Alamos needs this data, so can you give it to them?"

Wu had Havens sworn to secrecy, and agreed to release the draft paper. Nichols drove Wu to her apartment, and got the paper, which was already typed up. The draft paper was forwarded to Los Alamos.[20]

Wu's draft paper was a significant contribution to the smooth progress of the Manhattan Project. The present day nuclear reactors use a zirconium alloy to wrap around the uranium rods. The wrapping is to prevent xenon from leaking and slowing down the chain reactions.

The Manhattan Project presented a challenge to sciences, and more so to engineering and technologies. For example, the design and calculation of ignition and critical mass were complex and difficult. They required much scientific knowledge, complemented by engineering achievements.

People in charge were actually very worried when the test of the bomb was approaching. Oppenheimer was afraid that an unexpected explosion would be triggered in the air, and that there would be other unforeseen products.[21]

The first atomic bomb was successfully tested in a desert in New Mexico on July 16, 1945. Its trembling power, the blinding light, and the giant mushroom cloud signaled the arrival of a new era. Three weeks later, the two atomic bombs were dropped on Hiroshima and Nagasaki, effectively ending the Second World War.

The atomic bomb was a collective product of many scientists and engineers in the 20th century. Its destructive power shook the world, and the participating scientists felt guilty and responsible for its mass killings. But Germany had been conducting a similar development. The US scientists felt that if Nazi Germany won the race to develop the bomb, the end result would be even more disastrous.

After the successful test of the atomic bomb, there were different opinions as to whether it should be used in Japan. Scientists like Oppenheimer suggested dropping the bomb on an uninhabited island to demonstrate its power, but the final decision was to use it on a populated city.

Many participating scientists worked hard after the war to have the atomic weapon placed under the control of an international agency, and avoid its use in international conflicts. This led to the creation of the International Atomic Energy Agency (IAEA). Politicians would consider scientists naïve in view of the international political interests.

Oppenheimer was among the scientists championing international control of atomic energy. He was very disappointed when he learned in March 1946 that President Truman had appointed someone inappropriate to be the spokesman for the US delegates on international control of atomic energy in the United Nations.

D. Acheson, Chairman of the US Committee on International Control of Atomic Weapon Policy, arranged to have Oppenheimer meet with President Truman. Oppenheimer expressed his disappointment and, with a slip of the tongue, said: "Mr. President, I have blood on my hands."

Truman was very offended. He told Acheson later: "Never bring this guy to see me again. He just made the atomic bomb, but I was the one who decided to deploy it."[22]

Wu rarely talked about her chance involvement in the Manhattan Project. She said that it was painful learning about the defeats of Allies in Europe daily, and the suffering in China. It was certainly not easy for the US scientists to develop an atomic bomb in such a short time.[23]

When Wu was working at Columbia University, she once vacationed with the Swiss physicist W. Pauli in upstate New York. One afternoon, people rushed to get the news on the atomic bomb from the newspaper, but Wu and Pauli stayed put, as they had known the story already.[24] Switzerland was a neutral country, and Pauli was not involved in any defense research. He understood the suffering of Wu's family in wartime China, and her motivation in research on nuclear fission.

Some believed that Wu's contributions to China (through her contribution to the Manhattan Project) were immense and unmeasurable. Japan hastened its surrender, with the result that countless Chinese lives were spared in the battlefield.

Wu pursued research in pure physics after the war. She traveled twice to Taiwan, before visiting China for the first time in 1973. She returned to attend the meeting of Academia Sinica in 1962, and to receive the Special Contribution Award of the Chi-Tsin Culture Foundation in 1965. On the second trip to Taiwan, President Chiang Kai-Shek asked for her opinion whether Taiwan should develop an atomic bomb.

As an authority on the atomic bomb, Wu advised Chiang not to proceed, given the limited resources in personnel and finance in Taiwan.[25] China had been successful in producing an atomic bomb the previous year, and tested a hydrogen bomb two years afterward. Its own scientists and experts who had studied abroad in the Soviet Union and France accomplished these tasks.[26]

Wu had a sense of regret when asked about her involvement in the atomic bomb development. It was painful realizing its destructive power. She almost begged: "Do you think that people are so stupid and self-destructive? No. I have confidence in humankind. I believe we will one day live together peacefully."[27]

Notes

1. *From X-Rays to Quarks*, Emilio Segrè, p. 184, 1980.
2. Interview with Wu Chien-Shiung, March 1990, New York residence.
3. Interview with Luke Yuan, September 8, 1989, New York residence.
4. Wu Chien-Shiung, "Continuous X-Rays Excited by the Beta-Particles of $_{15}P^{32}$," *Phys. Rev.* 59, 481 (1941).
5. *Alvarez: Adventures of a Physicist*, Luis W. Alvarez, Basic Books, New York, 1987.
6. *What Little I Remember*, O. R. Frisch, p. 116, Cambridge University Press, 1979.
7. Interview with Wu Chien-Shiung, August 28, 1990, New York residence.

8. Diary of Glenn Seaborg, May 17, 1940. Provided by Seaborg.
9. Interview with Robert Wilson, September 21, 1989, Ithaca, New York residence.
10. Interview with Wu Chien-Shiung, August 28, 1990, New York residence.
11. *From X-Rays to Quarks*, Emilio Segrè, p. 260, 1980.
12. *Lawrence and His Laboratory: A History of the Lawrence Berkeley Laboratory*, J. L. Heilbron and Robert W. Seidel, Vol. 1, UC Berkeley Press, 1989.
13. Interview with Wu Chien-Shiung, September 18, 1989, New York residence. *Wu Chien-Shiung: The First Lady of Physics Research*, Gloria Lubkin, p. 52, Smithsonian, 1971.
14. Interview with Wu Chien-Shiung, September 18, 1989, New York residence. *Wu Chien-Shiung: The First Lady of Physics Research*, Gloria Lubkin, p. 52, Smithsonian, 1971.
15. Interview with Wu Chien-Shiung, September 18, 1989, New York residence.
16. *Shatter of Worlds*, Peter Goodchild and Robert J. Oppenheimer, p. 43, Fromm, New York, 1985.
17. Interview with William Havens, April 5, 1990, office in the American Institute of Physics, New York.
18. Interview with William Havens, April 5, 1990, office in the American Institute of Physics, New York.
19. *Enrico Fermi: Physicist*, Emilio Segrè, University of Chicago Press, Chicago, 1970.
20. Interview with Wu Chien-Shiung, September 18, 1989, New York residence.
21. Interview with Wu Chien-Shiung, September 18, 1989, New York residence.
22. *Shatter of Worlds*, Peter Goodchild and Robert J. Oppenheimer, p. 180, Fromm, New York, 1985.
23. Interview with Wu Chien-Shiung, February 12, 1990, New York residence.
24. Interview with Wu Chien-Shiung, July 31, 1990, New York residence.
25. Interview with Wu Chien-Shiung, March 4, 1990, New York residence.
26. The major contributors to the atomic bomb and hydrogen bomb developments in China included: Zhu Li-Ao, Qian San-Jiang (student of Irene Curie), Soviet-educated Wang Gan-Chang (student of Lise Meitner), and Peng Huan-Wu, Deng Jia-Xian, Yu Min, Zhou Guang-Zhao.
27. "1998 Research Corporation Award", interview in *New York Post*, January 22, 1959.

Chapter 8

A World Authority in Beta Decay

Wu Chien-Shiung went to Columbia University as a Senior Scientist in March 1944. She was delighted to resume her experimental research after almost two years of teaching at Smith College and then Princeton University.

At first she joined the wartime Manhattan Project in nuclear fission research, focusing on the development of gamma-ray detection.

When the Second World War ended in the second half of 1945, the wartime research also stopped. Wu was among just a few physicists who stayed behind at Columbia University. Given the prestige of Columbia's physics department, and the prejudice and discrimination of Ivy League schools against women, especially Asian women in the 1940s, her decision to stay was highly unusual.

When she came home from New York City on weekends she reviewed scientific periodicals in the library at Princeton University, which was open 24 hours. Wu prepared a concise survey and summary of theories and experiments on beta decay.

In preparing this review, Wu developed a better understanding of the history, development, status, and the critical problems in beta decay. Wu used this survey to select nuclear beta decay as the research area for her future experiments.[1]

The development of beta decay study started in 1895 when the German physicist Wilhelm C. Röntgen discovered X-rays. In 1901, he was awarded the first Nobel Prize for this discovery. 1895 was also important in modern Chinese history because China was defeated by Japan and signed the Treaty of Maguan with many concessions. Wu's father was only seven years old at that time and Wu was born seventeen years later.

X-rays were a major discovery. The French scientist Antoine Henri Becquerel, a contemporary of Röntgen, rushed into action. He accidentally discovered natural radioactivity from nuclei, which emit alpha rays (positively charged), beta rays (negatively charged), and neutral gamma rays.

Becquerel's understanding of radioactivity was supplemented by the research of two other French physicists — the famous Curie couple.

Many have argued that the better-known Madame Curie had benefited much from her husband, Pierre. They jointly discovered the radioactive elements radium and polonium, which greatly advanced the understanding of radioactivity. They shared the Nobel Prize in Physics with Becquerel in 1903. Marie Curie also won the Noble Prize in Chemistry, in 1911, for the discovery of radium and polonium, after the death of Pierre in an accident involving a horse-drawn carriage five years before.

Great advances were made from 1910 to 1930 in understanding the negatively charged rays in radioactive beta decay as classified by Madame Curie. The Austrian Lise Meitner, in particular, did very important work. She could be the most important figure in 20th century nuclear physics after Madame Curie. Being female and Jewish, Meitner was not properly recognized. Wu, herself an expert in this field, had the highest regard for Meitner, and even believed that Meitner's contributions were above those of Curie.[2]

Scientists at first used the images left on the photographic negatives to discover nuclear radiation. The images from beta decay were too fuzzy to be analyzed quantitatively. Only when Meitner and her collaborator Otto Hahn devised a new method with higher resolution between 1911 and 1912 did better experimental evidence for beta decay become available.

The complex spectra of beta decay were confusing to the scientists, and their proper interpretation was a major challenge. James Chadwick, who later discovered the neutron, confirmed experimentally in 1914 that the spectra of the electrons in beta decay are continuous and limited in energy. Eight years later, Meitner argued that a *quantized* (discrete) nucleus could not emit electrons of *continuous* energy.

The question was: Why are the spectra of the electron continuous in beta decay?

Most physicists (including Meitner) were leaning toward the idea that the electrons from beta decay possessed discrete energies initially, but switched to continuous energy later on due to other, unknown factors. C. D. Ellis was the only dissenter, arguing that the electrons had continuous energies from the very beginning.

The different opinions held by Meitner and Ellis led to a serious competition between the two. Their contemporary Wolfgang Pauli wrote Wu in 1958 telling her about the competition.

Pauli first admitted to being very competitive when he was young. He wrote: "I am so excited by the arguments between Meitner and Ellis (I love a good fight)."[3]

He wrote: "I first tried to talk to Meitner diplomatically without success. Then I admitted 'I think Ellis is right.' She turned red-faced with anger, but we had a long discussion nevertheless."

Ellis and W. A. Wooster did a beautiful experiment demonstrating continuous spectra of electrons in beta decay in 1927. Pauli reminisced in his letter to Wu: "When reminded of this result, our good Lise said 'I don't believe it! I will do a better experiment!'" Three years later, in 1930, Meitner and W. Orthmann published their new experimental result, which was similar to that of Ellis and W. A. Wooster. Pauli was very happy that they had pleasantly settled their argument through experiments.

Pauli also commented: "This experiment was Meitner's masterpiece." Her experiment also showed that only beta decay had continuous spectra, and the spectra in gamma decay were discrete. Wu referred to the letter from Pauli in a lecture in 1977, saying: "This incident illustrated clearly that healthy competition, and focus on experimental results, were deeply grounded in the early days of nuclear research."[4]

The critical experimental result of continuous energy spectra in beta decay created a big problem — lack of energy conservation — in nuclear physics. Besides, it was not clear if quantum mechanics was applicable at all in the domain of nuclear physics.

Bohr, who proposed an atomic model, said in a Faraday Lecture in 1930: "At this stage, we can safely say that there is no experimental nor theoretical evidence to support energy conservation in beta decay."[5] One reason for this puzzle was the belief then that electrons were *inside* the nuclei.

Pauli came to the rescue again. This sharp Austrian theoretical physicist suggested that there were *two* particles emitted in beta decay, based on considerations of the spin of the nuclei and statistical mechanics.

Wu had a strong bond with physicist Wolfgang Pauli.

Besides the electron (which was observed), there was another penetrating neutral particle (therefore not observed), with mass close to zero, which carried energy away from the electron (in a continuous way) and thus solved the energy conservation problem.

Surprisingly, Pauli never published the proposal, but only suggested this possibility in a letter to colleagues at a conference in 1930. He vaguely called the particle a neutral particle.

This neutral particle was introduced to conserve both energy and angular momentum (spin and statistics) in beta decay. The otherwise-present conflict in spin and statistics had challenged the original hypothesis that both protons and electrons were inside the nuclei.

There is a famous story regarding the confusion over whether electrons were inside the nuclei. W. Heisenberg, who was one of the founders of quantum mechanics and later a leader of the German atomic bomb project, was always skeptical of the hypothesis. On a hot day in the spring of 1931, he sat with the Austrian scientist Victor Weisskopf outside of a swimming pool. He asked: "We see these people fully dressed entering and leaving the pool house. Can you deduce that they swim with their clothes on?"[6]

In June 1931, Pauli attended the annual meeting of the American Physical Society in Pasadena. He again proposed the concept of this undetectable particle, and was greeted with general skepticism. He met Fermi in October of the same year at a nuclear physics conference in Rome, and they had a long discussion about beta decay. Fermi was intrigued by the proposal of a new neutral particle.

James Chadwick discovered the neutron in 1932. It was a breakthrough. The model of the nucleus made up of protons and neutrons was consistent with all previous developments, and presented an elegant picture. Pauli again proposed a new neutral particle in beta decay at the Solvay Conference the next year. Fermi was there and named the new particle a *neutrino* ("small neutral particle" in Italian).

Fermi vacationed in the Alps in northern Italy during the Christmas of 1933, together with Emilio Segrè, Edoardo Amaldi, and Franco Rasetti. In the hotel after skiing, he presented a new theory. It was the famous "Fermi's theory of beta decay". The paper was first submitted to *Nature*, but was rejected as it was too abstract and had no practical value. It was later published in *Nuovo Cimento* in Italy, and then in the *Zeitschrift für Physik* in Germany.

Fermi was a physicist of great depth and breadth. Some said that if he did not accomplish anything else, the 1934 *Nuovo Cimento* paper alone would place him as one of the top theoretical physicists of the 20th century.[7]

The great physicist C. N. Yang said in a lecture that Fermi was very solid and concrete in physics research. Many other physicists' work would suddenly become clear after his touch. Yang also praised Fermi as the last physicist to make first-rate achievements in both theories and experiments.[8]

Pauli never published his proposal of a neutrino in beta decay. Rabi (Chairman of the physics department of Columbia University) was interviewed by the American Institute of Physics for the oral history project in 1963. He recalled that Pauli, in his visit to the US, told him in a Chinese restaurant that he did not publish the neutrino proposal because "I consider myself more clever than Dirac." He referred to Dirac's proposal of the "hole" theory of antiparticles. Actually, scientists did not detect the elusive neutrino experimentally until the mid-1950s.

The research in neutrino was also related to another Chinese physicist, Wang Kan-Chang. Wang belonged to the senior generation of distinguished physicists in China, and was five years older than Wu. He studied with Meitner at the Kaiser Wilhelm Research Institute in Berlin, where he proposed investigating radioactive phenomena. Unfortunately, he did not get her support, and missed the opportunity of discovering the neutron.

Wang returned to China after four years in Europe. He diligently built up Zhejiang University, which was honored as "the University of Cambridge in the East" by Joseph Needham. He wrote a paper on the detection of the neutrino during the Sino-Japanese War. He pointed out that the only hope of detecting the neutrino was through the measurement of the (minute) recoil in the momentum of the (parent) radioactive element. The paper was published in the *Physical Review* in 1942. It was critical, direct, original, and was instrumental in the later experimental proof of the existence of the neutrino.[9]

While in China, Wang continued with his achievements in nuclear physics. He changed his name to Wang Jing, and joined the atomic bomb development effort. He was honored as the "father of the atomic bomb in China".

Some parts of physics research halted after the start of the Second World War, but the atomic weapon development during the war led to the nuclear reactor facility, which became a source of high intensity beta decay. A new magnetic spectrometer (also called a velocity spectrometer) was developed. Both benefited the further study of beta decay.

Wu's first experiment after the war was to repeat an earlier experiment by Luis Alvarez. Alvarez and another scientist had measured neutron scattering in hydrogen, and their results disagreed with theory. Wu and others repeated the experiment carefully using the new slow neutron magnetic spectrometer at Columbia. They found that the disagreement with theory was caused by a small defect in the earlier experimental measurements.

Wu did not have a teaching post at Columbia, but collaborated with several scientists as a Senior Scientist. These scientists included Rainwater, Havens and the wartime Project Manager Dunning. They coauthored a paper in the *Physical Review* in 1946 on their experimental results. Wu did most of the work.

Wu and others published several papers on slow neutron scattering studies of various elements. Fermi accidentally discovered the slow neutron effects in 1933, and created many artificial radioactive elements with his "Rome School" scientists. But they missed the discovery of the very important nuclear fission.

Besides slow neutron studies, Wu and others continued the development of high sensitivity radioactive particle detection. This work started in the wartime Manhattan Project. Its further development created a foundation for Wu's later achievements in beta decay.

After Fermi proposed his theory of beta decay, many experiments observed many more slow electrons than the theory predicted. George Uhlenbeck, who discovered electron spin, and Emil Konopinski, both of the University of Michigan, proposed a revised theory to resolve the discrepancy. This was later called the Uhlenbeck–Konopinski theory. As the effect of beta decay was very weak and very difficult to measure, the new theory was not able to resolve all of the discrepancies either. The situation was far from being settled.

A slow electron is like a slow moving golf ball. The movement of the ball is affected by minute imperfections on the golf course. Similarly, the scattering of slow electrons is very sensitive to the detector's environment. Because Wu had been working on beta decay, she had an in-depth understanding of detection techniques. She was naturally interested in this discrepancy between theory and experiment.

In 1949, Wu and her collaborator R. Albert performed an experiment to investigate the slow-electron discrepancy in beta decay. Wu reasoned that if the slow electrons were sensitive to the environment, they must be affected by the thickness and uniformity of the radioactive source. She devised a very simple but very clever technique to prepare a thin and uniform source.

She introduced drops of a chemical solution like detergent into a bucket of water. The solution would spread to form a thin film. She used a copper ring to scoop up a round film, and placed a drop of solution containing radioactive copper on the film. The radioactive copper spread out uniformly and very thinly, due to surface tension. The round film would be used as source when dried.[10]

This simple idea was very difficult to execute. Many attempts, including those by Fermi, were unsuccessful. Wu's experimental result of beta ray spectra was in perfect agreement with Fermi's theory. This paper firmly settled the arguments about beta decay. Wu had established her position in beta decay as a world-renowned, ultraprecise experimental physicist.

Wu continued to work on beta decay after 1949, and discovered several new beta ray spectra. In nuclear physics, the earlier beta ray spectra were called "allowed" spectra, and the later ones "forbidden" spectra. Wu investigated the spin and parity states of these forbidden spectra, and identified three different classes. As she wrote in a paper: "When it rains, it pours."[11] The experimental observations of these first-, second-, and third-order forbidden spectra all agreed with theoretical prediction. Fermi's theory of beta decay was further advanced.

In the early 1950s, Wu had become the authority on beta decay, which was an important, highly difficult and complex field. She was not yet 40 years old. Her university title was "associate", which allowed her to advise graduate students. To make sure that the university administration did not overlook her, her collaborator Havens arranged to have her give several lectures. Her laboratory, in the basement of the Pupin Laboratories, was moved to the thirteenth floor only a few years after she had been promoted to associate professor in 1952.

Before 1951, her colleagues at Columbia discussed several times if they should offer her a teaching position, but without success. Wu was totally devoted to her research, and had no interest in fighting for promotion or salary.[12]

Wu in the laboratory of Columbia University.

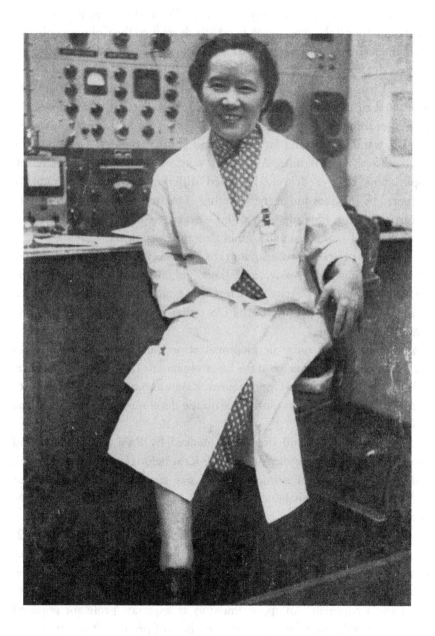

Wu's friend Ursula Schaefer from the UC Berkeley days, and her husband, Willis Lamb, were both at Columbia University. Ursula was a history professor at Barnard Women's College, and Lamb was a physics professor at Columbia. In 1951, Lamb proposed that Wu be appointed to a teaching position. Isidor Rabi, who single-handedly developed the Columbia physics department into a world-class institution, turned down the proposal. The only reason was that Wu was a woman.[13]

Rabi was born in a small town in the Ottoman Empire in 1898. He emigrated to the US and grew up in New York City. He maintained the Jewish tradition, which values education. Rabi went to Cornell University in 1916, wandered around for a while, and graduated in 1919 with a major in chemistry. At that time, Americans were more interested in practical applications than a degree. He tried different things, was gradually intrigued by the question of the structure of matter, and slowly returned to research. After graduating with a Ph.D. degree in 1927, he toured Europe for a couple of years, like many other American scientists.

Rabi was offered a professorship at Columbia in 1929. He was a distinguished physicist, progressive, ambitious, and far-sighted. He soon became very influential. The physics department of Columbia University emerged as the top research institute in the US under his leadership. He won the Nobel Prize in 1944.

Rabi joined the radar development effort during the Second World War, as an Associate Director of the Electromagnetic Radiation Laboratories at MIT, and an Advisor to Oppenheimer. Rabi was more involved in politics after the war, and had great influence on the development of science in the US and the world.

Wu was close to and greatly influenced by Rabi at Columbia and Segrè at Berkeley in her science career. Rabi belonged to the old school, and had his opinion of women. But he was genuinely concerned about Wu. Until his death on January 11, 1988, he maintained this close relationship with her.

Wu was finally promoted to associate professor with tenure in 1952. This was highly unusual at Columbia. The physics department was a premier institute in the US. Besides Rabi, there were the future Nobel laureates Lamb, Charles H. Townes, James Rainwater, Jack Steinberger, and Leon Lederman. An appointment as an associate professor required the recommendation of an external committee of prominent scientists.

Although Wu became an associate professor, her salary was still low. It stayed that way until the new chairman, Robert Serber, noticed the discrepancy and made a drastic adjustment in 1975.[14] Wu never publicly discussed this matter; she devoted her attention to research.

Wu taught a course in nuclear physics, and had several graduate students. She was not the best lecturer when compared to those famous physicists at Columbia University.[15] But she was a pioneer in nuclear physics, and her course materials often contained her own research, with very solid and profound insights. Her student Noemie Koller, Associate Dean for Sciences at Rutgers University (1992–1996) in New Jersey, said that it was the best course she had ever taken.[16]

There was no good textbook in nuclear physics back then. Wu invited the theoretical physicist Steven Moszkowski of UCLA to coauthor a textbook called *Beta Decay*. Because of their busy schedules, the textbook was not published until 1966. It became a classic, and was highly regarded by students and researchers.

Lee Grodzin of MIT, a colleague of Wu, wrote her after the publication; he said that the book was so well written and informative that he could not put it down, and read it from cover to cover. Koller also said that this textbook was a must for any physics student. Whoever had trouble understanding it had better do something else than physics as a profession.[17]

Wu behaved like a stern parent and treated her students like children. She had high expectations of them, but also gave them a lot to do. She was not like a famous professor on a high pedestal looking down on the students; rather, she always chatted with them about physics or other issues. Her students said that their research group was like a close family, working all day long side by side, frequently having lunches and dinners together. A student who worked with her closely recalled that when he and his girlfriend once met Wu on the street, she stopped and waited for them, making sure that he would introduce the girl.[18]

Wu worked very hard. She typically started around eight in the morning, put in a long day teaching and doing research, and stayed in the laboratory until seven or eight in the evening, with lunches and dinners in either the university cafeteria or neighborhood restaurants. Sometimes she would stay until midnight. Koller said that Wu frequently carried home technical problems after leaving the laboratory at night, and had them solved the next morning.

Wu believed in total devotion to research. She practiced this belief, and expected the same of her students. She could not understand how a researcher could be distracted. She demanded perfection from her students. The most precise measurement, and accurate calculation, in every step of every experiment. She asked her students to work all day on weekdays, Saturdays, and sometimes Sundays. She did not shy away from expressing her disappointment when her students failed to meet her demands.

A big sign was hung outside her laboratory in Pupin that read "No Radioactive Materials Permitted Beyond Line".

In the 20th century, research in science had become much more competitive for funding support and recognition. As such, it was a far cry from the earlier cooperative atmosphere. People who knew her well believed that she was highly competitive by nature, and she felt that she had to double her effort in order to succeed, given what she had gone through in a rather oppressive environment.

Wu's students recalled that she was the undisputed authority in the laboratory, and that she would never hesitate at all to correct any mistake that students had made. She preferred to lead a team of students, rather than collaborating with other scientists. She had very good insight, and always selected the most difficult — and the most fundamental — problems to attack.[19] Students would argue with her about physics problems, but her opinions always turned out to be right. The students later found out that her opinions and understanding were results of years of experience. They very much admired her insight into physics.[20]

As her experimental work faced increasing competition, Wu told her students not to show their data to visitors until it was published because it might be stolen. When guests pried, she switched to a particularly convoluted form of Chinese–English. She could talk and talk in a charmingly soft voice and yet somehow never answer the question. Wu was fundamentally a serious and reserved person, not that open and straightforward. In a way, she was a rather private person.[21]

One of her students remembered her rather defensive personality. There was a famous European scientist visiting Columbia. His English was not that fluent, and he would occasionally ask students for the right word during his lecture. By contrast, Wu could never be comfortable without having prepared perfect lecture notes in advance.[22]

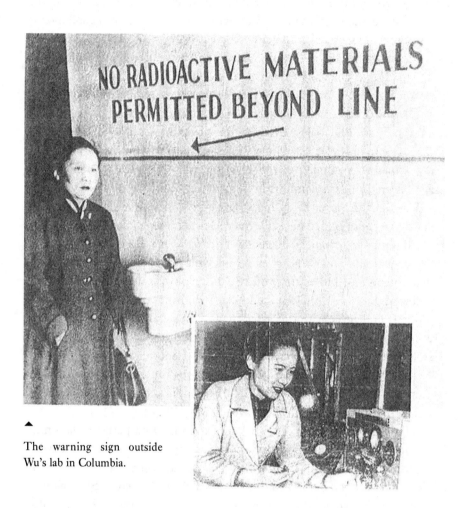

The warning sign outside Wu's lab in Columbia.

Wu in the laboratory.

Sometimes, Wu would become really angry at the students. She would never yell at them, but would quietly go into her office, not talking to the students. Around that time, the students considered installing a light beam detector at a certain height to monitor her mood. If she was in a good mood, she would be walking tall with her head up, thus triggering the light beam.[23]

The students did not really install the monitor. But they gradually understood, and got used to her quiet stubbornness. Wu was nicknamed "Dragon Lady" in Colombia's physics department in the early 1950s.

"Dragon Lady" was the name given to a glamorous but dangerous Chinese beauty. She was a character in a popular US newspaper comic strip, *Terry and the Pirates,* telling stories set near Hong Kong. Some said that the character alluded to the very powerful First Lady Soong May-Ling. The artist of the comic strip was Milton Caniff.[24]

Wu's signature *qipao* dress, her strong opinions, and her demands on the students naturally landed her with the nickname. Her students occasionally referred to her by this nickname behind her back, but did so with affection. Her students and contemporaries recalled that she was actually the most humane, generous, and warm professor in the physics department. Most of the others were rather self-centered folks, with little concern for students.[25]

One would agree with this assessment with some understanding of the developmental history of the physics department of Columbia University. Besides Rabi, a string of Nobel laureates like Lederman, Steinberger, and T. D. Lee (who arrived in 1953) were notorious for their strong egos and toughness. After Lederman and Steinberger left, Lee became the sole power in the 1970s.

Before Wu started her earthshaking experiment on parity violation in 1956, she continued to focus on research in beta decay. Beta decay was very complicated, with many sometimes conflicting experimental results, which made it difficult to be consolidated into a general theory. She made a regrettable mistake in the process of solving this puzzle. A particular experiment was done by her students, and her name was not on the paper. But people never forgot the mistake as it had come out from Wu's laboratory, which was well known for its standards of precision and accuracy.

As the saying goes among physicists, "People may not remember all your other experiments with correct results, but will surely remember your single mistake." This reminds one that the human error factor has a role in scientific research too.

The theory of beta decay can be generalized in the following way: the matrix elements describing the transition of nuclear states can be coupled to the electron–neutrino field as a scalar (S), vector (V), tensor (T), axial vector (A), or pseudoscalar (P). The experimental challenge was to distinguish the form of the transition and the relative strengths.

For example, the matrix element for the emission of a photon (quantum field), while the source undergoes a state transition, is called a vector (V) in quantum electrodynamics. The vector quantum field has spin 1 and negative parity. Similarly, an axial vector (A) quantum field has spin 1 and positive parity, and a scalar (S) quantum field has spin 0 and positive parity.

The original Fermi theory of beta decay was proposed by analogy with quantum electrodynamics with a vector (V) matrix element. Two theoretical physicists, George Gamow and Edward Teller, later proposed another beta decay theory, the Gamow–Teller theory, allowing transitions with a change of spin. Both theories and the allowed transitions were confirmed. The physicists believed then that a scalar (S) or vector (V) matrix element gives rise to the Fermi theory (the selection rule for allowed transitions), whereas a tensor (T) or axial vector (A) matrix element gives rise to the Gamow–Teller theory (the selection rule for allowed transitions).

In the early 1950s, two of Wu's students, S. Ruby and B. Rustad, performed an experiment to investigate the beta decay in the transition from radioactive helium (He-6) to lithium (Li-6).

Wu held discussions with the students during the experimental process. The students published a short article in *Physical Review Letters* in 1952, followed by a long article in the *Physical Review* in 1955. They determined that the Fermi theory had a scalar (S) transition matrix, and the Gamow–Teller theory had a tensor (T) transition matrix.

As their experiment had Wu's endorsement, and she had a long record of precision, the Ruby–Rustad papers initially carried a lot of credibility. Later experiments, however, showed conflicting results.

Richard Feynman, M. Gell-Mann (who won the Nobel Prize for a proposal of "quarks" and their interactions.), R. Marshak and his student E. Sudarshan, and another physicist, J. Sakurai, all argued that the transition matrices in beta decay were vector (V) and axial vector (A). Before this was settled, some said that Marshak must be mad. How could the He-6 experiment be wrong?

Not long afterward, Maurice Goldhaber and two collaborators did an elegant experiment and proved that the $V-A$ theory was correct. That settled the dispute.

Wu was very unhappy about the mistake made in the experiment of Ruby and Rustad. Ruby discussed the experiment in the Plaza Hotel (a landmark in New York City where Chiang Ching-Kuo, then Vice-Premier and later President of Taiwan, was shot while visiting the US) in January 1990, and regretted that he was so careless. He did not finish his Ph.D. degree, worked for IBM for some time, and resumed research work at Stanford University. Rustad died in the early 1960s.

The incident bothered Wu. She later built a larger experimental setup at Columbia, and did a similar experiment with He-6. She and her collaborator Arthur Schwarzchild wrote a paper in 1958 pointing out the factors causing the mistake in the earlier experiment.

This bad mark did not change very much the position of authority in the field of beta decay that Wu enjoyed. Her reputation as the most precise experimentalist was intact. The saying in the physics circle was: "If the experiment was done by Wu, it must be correct."[26]

Wu's interest was always in beta decay; when she tried her hand at particle physics while visiting Los Alamos one summer in the mid-1950s, she was not excited about it.[27] She had a special love for beta decay, and once said: "Beta decay is like a dear old friend. There will always be a place in my heart especially reserved for beta decay."[28] She believed that beta decay was full of surprise and delicacy.[29]

In the mid-1950s, Wu was a first-rate explorer in this field of research full of surprise and delicacy.

Notes

1. "A Few Highlights of C. S. Wu's Research Contributions to Physics", Wu Chien-Shiung, and "Public Talk of Wu" while visiting Nanjing University, May 21, 1994.
2. Interview with Wu Chien-Shiung, June 21, 1990, New York residence.
3. "Surprise and Delicacy — Contribution of Beta Decay in the Understanding of Weak Interaction", Wu Chien-Shiung, at "Fifty Years in Weak Interaction" International Congress, January 1977.
4. "Surprise and Delicacy — Contribution of Beta Decay in the Understanding of Weak Interaction", Wu Chien-Shiung, at "Fifty Years in Weak Interaction" International Congress, January 1977.

5. "Surprise and Delicacy — Contribution of Beta Decay in the Understanding of Weak Interaction", Wu Chien-Shiung, at "Fifty Years in Weak Interaction" International Congress, January 1977.

6. Interview with Victor Weisskopf in the Oral History Project, July 10, 1963, AIP archive, New York.

7. *The Tenth Dimension*, Jeremy Bernstein, 1989. McGraw-Hill, New York.

8. Lecture by C. N. Yang delivered in June 1986, Chinese Science and Technology University.

9. "Wang Kan-Chang and the Development of Neutrino", Li Bing-An and C. N. Yang, 15, 12, *Physics*, 1986.

10. Interview with Mo Wei, March 24, 1990, Blacksburg, Virginia residence.

11. "Trends in Atomic Physics", C. S. Wu, in *Essays Dedicated to Lise Meitner, Otto Hahn, Max von Laue on the Occasion of Their 80th Birthday*, Ed. O. R. Frisch *et al.*, 1959, Interscience, New York.

12. Interview with Noemie Koller, Charles Townes and others.

13. Private letter to the author from Willis Lamb, March 27, 1994.

14. Interview with Robert Serber, March 14, 1990, New York residence.

15. Interview with Evelyn Hu, April 1, 1990, office in UC Santa Barbara.

16. Interview with Noemie Koller, December 12, 1989, office in Rutgers University.

17. Interview with Wu Chien-Shiung, January 27, 1990, New York residence.

18. Interview with Noemie Koller, December 12, 1989, office in Rutgers University. Interview with Mo Wei, March 24, 1990, Blacksburg, Virginia residence.

19. Interview with Felix Boehm, October 20, 1989, office in Caltech.

20. Interview with Noemie Koller, December 12, 1989, office in Rutgers University.

21. Interview with Noemie Koller, December 12, 1989, office in Rutgers University. Interview with Felix Boehm, October 20, 1989, office in Caltech.

22. Interview with Stanley Ruby, January 15, 1990, Plaza Hotel, New York.

23. Interview with Noemie Koller, December 12, 1989, office in Rutgers University.

24. Interview with Walter Sullivan, science reporter of *The New York Times*, December 15, 1989, New York.

25. Interview with Stanley Ruby, January 15, 1990, Plaza Hotel, New York. Interview with Noemie Koller, December 12, 1989, office in Rutgers University.

26. Interview with H. Schopper, August 6, 1989, CERN, Geneva.

27. Interview with Wu Chien-Shiung, February 12, 1990, New York residence.

28. "The Discovery of Non-Conservation of Parity", Wu Chien-Shiung, in *Beta Decay in Thirty Years Since Parity Non-Conservation: A Symposium for T. D. Lee*, ed. Robert Novick, 1988, Birkhauser, Boston.

29. "Trends in Atomic Physics", C. S. Wu, in *Essays Dedicated to Lise Meitner, Otto Hahn, Max von Laue on the Occasion of Their 80th Birthday*, ed. O. R. Frisch *et al.*, 1959, Interscience, New York.

Chapter 9

Revolution in Parity Conservation

Heavy snow fell in Washington, DC on December 24, 1956. Dulles and Washington National airport were closed, and the commuters between New York City and Washington rushed to the Union Station to catch the train.

A petite, middle-aged, Asian woman was among the commuters. She bought a ticket for the last train to New York.[1] She did not get any attention from fellow passengers.

The lady was Wu Chien-Shiung. This trip had a special meaning for the history of science, as the experimental results that she brought home would create a revolutionary change in 20th-century physics.

Her results made 1957 a special year for the Chinese in the development of modern science. That year, two Chinese physicists won the coveted Nobel Prize in Physics. To date, they remain the only winners with Chinese nationality.[2] These two Chinese are C. N. Yang, Einstein Professor at SUNY (State University of New York), Stony Brook, and T. D. Lee, a professor at Columbia University. They boldly questioned a long-held concept in physics and suggested experiments to test their hypothesis. The hypothesis was first proven by Wu's undisputed experiment, and they won the prize in record time.

Although the Lee–Yang hypothesis had only been made the previous year, the breakthrough occurred at the end of a long period of scientific progress.

Searching for the basic structure of matter has always been a tradition of Western science. From the very beginning in Greece, it was believed that matter was made up of "atoms", "indivisible" in Greek.

The idea that atoms were the basic "indivisible" units was held firm, until the discovery of nuclei inside the atoms by the British scientist E. Rutherford at Manchester University in 1911.

Scientists then discovered a positively charged *proton* and a negatively charged *electron* inside the atoms. Protons and electrons were first believed to be *inside* the nuclei, but the assumption proved inconsistent with observations. The British scientist J. Chadwick discovered a neutral particle, the *neutron* — in 1932, and confirmed that protons and neutrons were inside the nuclei, with electrons circulating outside. In the same year, C. D. Anderson of Caltech discovered a new particle — the *positron* — in cosmic rays.

The positron is the electron's antimatter. It has the same mass but is positively charged. Einstein proposed a particle, the *photon*, to explain the discrete characteristic of light (photoelectric effect) in 1905. By the end of 1932, scientists knew of five "fundamental" particles.

The number of fundamental particles increased to forty or so in the 1960s, the exact number depending on what one defines as "fundamental". Currently, the "fundamental" particles are believed to consist of *leptons*, *quarks*, and *gauge bosons*. Leptons and quarks each have three varieties (triplets), and there are four gauge bosons mediating the four forces (gravitational, weak, electromagnetic, and strong). The theory of the structure and interactions among these particles is called the *Standard Model*.

As stars burn in the universe, they emit high-energy cosmic rays. After their first detection from the top of the Eiffel Tower in Paris in 1910, scientists started the detection and measurement of cosmic rays from high mountains — the Alps of France, the Rockies of America, the Andes of South America — and from the top of skyscrapers. They even put detection apparatus in balloons and airplanes. When Luke Yuan conducted research at Princeton University in the 1940s, he also set up apparatus to investigate cosmic rays at the top of the Empire State Building, then the tallest Building in the US.[3]

Scientists used a "cloud chamber" to detect cosmic rays on the ground. An energetic charged cosmic ray passing through the supersaturated vapor in the chamber knocks electrons out of the atoms in its path and leaves a track of ions in its wake. The condensation of cloud droplets on these ions takes place all along this path, and thus the track of the ionizing cosmic ray can be seen and photographed. Anderson discovered the positron in this way. For a long time, the cosmic ray was the only source for studying new particles with very short half-lives.

Cosmic rays travel a long distance across the universe, and are absorbed and deflected by the atmosphere and the magnetic field of the earth. Their energy and intensity cannot be easily controlled.

Then came the concept of an accelerator — a machine to produce a laboratory-accelerated particle beam. In 1932, two British scientists, John Cockcroft and Ernest Walton, constructed what was called a Cockcroft–Walton accelerator — a linear accelerator using a high-voltage ac source to accelerate protons to 600,000 electron volts. Lawrence succeeded in substituting a circular racetrack for the straight-line track of the linear accelerator, and a succession of small pushes for a few large ones, thus requiring smaller voltages.

The first *cyclotron*, barely one foot across, accelerated protons to 1,200,000 electron volts. It drastically changed the landscape of particle research, and earned Lawrence the 1939 Nobel Prize in physics. Two larger cyclotrons were built in the 1950s. They provided new tools for particle physics experiments, and a backdrop for the Lee–Yang hypothesis.

The *Cosmotron* — a machine to produce charged particles with energy as high as that in cosmic rays — was completed at the Brookhaven National Laboratory in 1953. Another accelerator, called the *Bevatron*, was completed at Berkeley in 1955.

Before the last two accelerators were completed, scientists had already discovered many new particles in cosmic rays. These were named "strange particles", as none of them were predicted by theory.

Two British experimental physicists, George Dixon Rochester and Clifford Charles Butler, first discovered these strange particles when observing cosmic ray effects in a cloud chamber in 1947. The particles were quite different from ordinary matter.

Ordinary matter is composed of protons, neutrons, and electrons. When high-energy protons bombard matter, the resultant fragments contain strange particles. Two of these strange particles, θ(theta) and τ(tau), attracted special interest from physicists. They exhibited a strange and puzzling characteristic — the so-called θ–τ *paradox*.

θ and τ are produced in the fragments created when ordinary matter is bombarded by cosmic ray or high-energy particles in accelerators. They have short lives, and readily decay into other, longer-lived particles which physicists can observe.

Physicists were puzzled by the $\theta-\tau$ *paradox* as the θ decays into two π mesons (pions), and the τ decays into three π mesons. The Japanese physicist Hideki Yukawa had predicted in 1934 the existence of the meson — for which, after experimental confirmation, he was awarded the Nobel Prize in Physics in 1949. The exchange of pions gives rise to nuclear forces.

R. H. Dalitz demonstrated that θ and τ were unlikely to be the same particle based on accepted principles in physics as we knew then, yet they had a very similar mass and lifetime.

These conflicting results were at the heart of the $\theta-\tau$ paradox. Measurements of their mass and mean life were initially inaccurate, and physicists were leaning toward believing that they were indeed different particles. This solution to the paradox was all too easy.

Because they produced more particles and had better control of the energy, accelerators were useful for accurate measurements of the mass and lifetime of the θ and τ. In the second half of 1956, 60% of the machine time of the Cosmotron at Brookhaven was allotted to research projects investigating strange particles.[4]

By 1957, the measurements of the mass and lifetime of θ and τ were essentially identical within error bars. The two particles are now called K-mesons.

It was not right for the θ and τ to be identical particles with two modes of decay. Theoretical physicists put forward many different theories trying to solve this puzzle. They usually discussed their emerging new ideas in private conversations, letters or by telephone. However, the public discussion of the $\theta-\tau$ *paradox* reached a climax at the Rochester Conference in April 1956.

The Rochester Conference was an important annual international conference in particle physics. The 1956 conference was the sixth one, held at Rochester University in New York State, from April 3 to 7. On the last day of the meeting, there was a panel discussion on "Theoretical Interpretation of New Particles", chaired by Robert Oppenheimer. C. N. Yang first summarized and reported on different theories of strange particles, and presented his own and others' ideas. Murray Gell-Mann (1969 Nobel laureate) and Richard Feynman (1965 Nobel laureate) also presented their ideas and questions. The proceedings of that Rochester Conference recorded:

> Further discussion.... Yang believes that we should keep an open mind, as we know so little of the $\theta-\tau$ paradox. Following this line of thinking, Feynman asks a question of M. Block: Is it possible that θ and τ are the

same particle in two different modes of parity? As they have no specific parity, it means that parity is not conserved. Nature may not have a way to decide right-handedness or left-handedness. Yang reports that he and T. D. Lee had investigated this problem but have no conclusion yet.[5]

What is "parity" in this discussion?

Simply speaking, parity is left–right symmetry. Symmetry was an early concept. Physicists in the 20th century understood that momentum-energy conservation was a result of space–time symmetry — the momentum and energy of a physical system will not change before and after a physical inter-action, as the physical interaction is identical under space and time inversion. This principle of symmetry-to-conservation in Newtonian mechanics led to the concept of parity conservation in quantum mechanics.

Parity conservation stated that physical principles do not distinguish between left and right at the deepest level. It was also called "mirror sym-metry". If the mirror image of a physical phenomenon represents another possible physical situation, the phenomenon is said to conserve parity.

Consider opening a bottle with a clockwise motion while standing in front of a mirror. The image would show a counterclockwise motion. We say that the physical action (opening a bottle) is the same in the two worlds, and therefore has parity conservation.

Parity conservation was one of the principles that physicists believed in. It was very unusual to question the validity of this fundamental belief. Although the experiments involving strange particles raised the possibility of parity nonconservation, not many physicists took it seriously.

Yang recalled that space–time symmetry was so useful and valuable in molecular, atomic, and nuclear physics that people assumed it to be a first principle. Also, the principle of parity conservation was used quite success-fully in nuclear physics and beta decay. Any suggestion of its violation would meet with violent objection.

Yang believed that the breakthrough was to separate parity conservation in weak interaction from parity conservation in strong interaction. Otherwise, one would encounter conceptual as well as experimental difficulties.[6]

Yang and Lee continued to search for solutions to the θ–τ paradox after the Rochester Conference. Yang was at the Institute for Advanced Study in Princeton, which was directed by Oppenheimer, and after April, he visited the Brookhaven National Laboratory in Long Island for the summer. He continued to meet twice a week with Lee, who was at Columbia. The distance between Columbia and Princeton was about the same as that between Columbia and Brookhaven. Sometimes Yang visited Columbia, and sometimes Lee visited Princeton or Brookhaven. The revolutionary idea of parity nonconservation emerged at a meeting in May, when Yang visited Lee.

Yang and Lee both have written record of this process.[7] But their accounts differ on who originated the idea of challenging the principle of parity conservation. This difference caused their breakup in later years.

They agree that it was on a day in either late April or early May of 1956. Yang drove to Columbia University from the Brookhaven National Laboratory. They were planning to have lunch in a Chinese restaurant at West 125th Street & Broadway. As the restaurant was not yet open, they parked the car in front, and walked to the White Rose Café to continue their discussion, then moved back to the restaurant for lunch. They returned to Lee's office at Columbia, and had an enthusiastic discussion for the whole afternoon. The breakthrough from this Lee–Yang discussion was to focus the question of parity conservation on its validity in weak interactions.

In retrospect, this idea was quite trivial, but back then it was not that simple. Yang recalled his thinking: one had no idea where to go when facing a problem like the $\theta - \tau$ paradox. But once one found a clue, he could gather all his resources to solve the problem, rather than wandering around.[8]

Physics is an experimental science. Physicists can propose fancy theories, but these have to be confirmed experimentally. The Lee–Yang challenge of parity conservation in weak interactions could not have been confirmed so rapidly without Wu, who was an authority on weak interaction experiments, and had the determination to resolve such an important problem.

Wu joined Columbia University in 1944, working on the Manhattan Project. She focused on beta decay research after the war in 1945, and became an authority even before she was promoted to associate professor in 1952.

As beta decay was an important constituent of the weak interactions, and the Lee–Yang challenge of parity conservation zeroed in on weak interactions, they naturally considered seeking advice from Wu. In May, Lee went up from his own office on the eighth floor to his colleague Wu's office on the thirteenth floor in Pupin Laboratories.

Wu was not a particle physicist, and was not familiar with the θ-τ paradox. Lee explained the paradox, and his many discussions with Yang leading to their suspicion that parity may not be conserved in weak interactions. Wu, because of her profound understanding of beta decay and weak interactions, was immediately interested in this problem.

Lee was not that familiar with beta decay experiments in weak interactions. Wu then lent him a copy of the thousand-page book *Beta and Gamma Spectra*, by the Swedish physicist Kai Siegbahn. She remembered clearly that her own copy had been lent out, and she had to take someone else's copy to lend to Lee.[9]

Separately, Yang and Lee were busily doing calculations at Brookhaven and Columbia, respectively. After a couple of weeks, they discovered that the theoretical calculations of beta decay (which had much experimental confirmation) did not depend on the assumption of parity conservation.[10] They could not find any experimental evidence that parity was conserved in weak interactions. They were more convinced than ever to further investigate parity conservation in weak interactions, and prepared to write a paper on this idea.

Lee and Yang met with Maurice Goldhaber at Brookhaven one day. Goldhaber was trained in the Cavendish Laboratory at the University of Cambridge with Rutherford, and later became the Director of the Brookhaven National Laboratory. He recalled telling Lee and Yang that scientists at Oxford University had succeeded in polarizing the nuclei,[11] a technique that was mentioned in the Lee–Yang paper as being central to investigating the question of parity conservation.

Wu, an expert in nuclear research, had already been interested during the previous few years in the new technique of polarizing nuclei in the low-temperature laboratories at Oxford University and in Rotterdam, Holland. Simply speaking, to polarize a nucleus is to orient the nuclear spin in a certain direction with a strong magnetic field.

In Lee's next discussion with Wu, she asked if anyone was going to test his hypothesis. He mentioned Goldhaber's suggestion of using the nuclear polarization technique. She immediately suggested using polarized cobalt-60 as the source of beta decay as a test.

After those discussions, Lee and Yang finished writing a paper in June, with the title "Is Parity Conserved in Weak Interactions?". The paper raised the theoretical question whether parity is conserved in weak interactions, predicted theoretically several measurable consequences of parity violation in beta decay, and suggested experiments to test the validity of the parity conservation principle. Lee and Yang acknowledged five physicists, including Goldhaber and Wu, at the end. The paper was sent to the *Physical Review* on June 22, and published on October 1.

But the editors of the *Physical Review* did not allow titles ending with a question mark and the title was changed to "Question of Parity Conservation in Weak Interactions". The paper would earn Lee and Yang the Nobel Prize in Physics the following year, and become a classic in the scientific literature.

Before the Lee–Yang paper was completed, Wu had already realized that it was a golden opportunity for a beta decay physicist to perform a critical test. She believed that this would offer a tremendous contribution and provide experimental evidence, even if the results were positive — of whether parity is conserved in beta decay.[12] Yang recalled that they had talked to other physicists, but only Wu understood the importance. This showed that Wu was a distinguished scientist, as a distinguished scientist must have very good insight.[13]

Actually, the Lee–Yang paper did not claim that parity is not conserved for sure. It just proposed a possible direction for solving the $\theta - \tau$ paradox. Yang said later: "I know of nobody at that time, in the summer of 1956 — and that includes Lee and Yang — who believed that it would not be symmetrical. It was not clear to me how Telegdi thought.[14] We wrote our paper only because we thought it should be tested." He asserted: "Nobody believed it would happen and, because it was so difficult, they wouldn't tackle it. But Wu had the perception that right–left symmetry was so basic and fundamental that it should be tested. Even if the experiment had showed it was symmetrical, it would still have been a most important experiment. There was no past evidence of left–right symmetry in beta decay."[15]

Many experimentalists did not join the race, as the experiment was indeed very difficult. Because of her in-depth understanding of experimental techniques, Wu fully appreciated the difficulty. She said that Yang and Lee got the idea of an experimental test from her. She knew that the experiment would face two hitherto unmet challenges in nuclear physics: one was to keep the detectors (recording electrons coming out of beta decays) functioning properly at extremely low temperatures, and the other was to maintain the very thin (beta decays) source in a polarized state long enough to obtain enough decay event statistics.[16] Even with these foreseeable difficulties, but an uncertain outcome, she decided to proceed immediately.

Wu and Luke Yuan had planned a long trip that spring. They would first attend a high-energy physics conference in Geneva, and then conduct a lecture tour in Asia. That would be their first trip to Asia in twenty-plus years since they left China in 1936. As mainland China was closed behind the "bamboo curtain", they planned to visit Taiwan.

They had even booked a passage on the ocean liner *Queen Elizabeth*, planning to travel across the Atlantic Ocean. Wu realized the importance of this experiment and the need to start a clear investigation immediately. She asked Luke to go ahead, and leave her behind to do the experiment. Being a physicist himself, Luke understood the significance of this experiment, and went on the trip by himself. He attended the high-energy physics conference in Geneva, stayed briefly in England, France, Italy, and Egypt, visited and lectured at the Tata Institute in India, and arrived in Taiwan in July.[17]

Wu had done a lot of preparation for her forthcoming experiment during this time. She studied the most recent literature to understand the latest developments in nuclear physics using cobalt-60. As the experiment would involve both nuclear physics and low-temperature physics, she aggressively pursued the latest techniques involving the latter.

Low-temperature physics has a long history. In Europe, there were two low-temperature physics laboratories with long traditions. The pioneering physicist H. Kamerlingh Onnes established one in Rotterdam, The Netherlands. Superconductivity was discovered there. F. Lindemann established the other one: the Clarendon Laboratory at Oxford University. He emigrated to England from Germany, and served as Scientific Advisor to Prime Minister Winston Churchill during the Second World War. Nuclear polarization has to be done at very low temperature, to keep the nuclei from random movement. The technique was first developed in these two laboratories, in the early 1950s.

Wu herself was not a low-temperature physicist. She realized that she had to find a collaborator who was a good low-temperature physicist with a clear understanding of nuclear polarization. Columbia University had a low-temperature physics group, but its scope and equipment were not sufficient to meet these requirements. There was a physicist named Richard Garwin, working for IBM in a branch of the Watson Laboratory at Columbia. He was a fellow graduate student of Yang and Lee at the University of Chicago under Fermi, and was honored by Fermi as "the only real genius I have ever met".[18]

Garwin had earlier contributed to the development of the hydrogen bomb in the US, and later served as the Director of the IBM Watson Laboratory. He did research on low-temperature physics in his developmental work on high-speed computers using superconductors.

At first, Wu contacted Garwin and invited him to collaborate on the experiment to investigate parity conservation. He was busy starting a research project at IBM and declined.[19]

The National Bureau of Standards (NBS) in Washington DC was another US laboratory capable of doing nuclear polarization experiments at low temperature. Wu knew that Ernest Ambler there had come from the Clarendon Laboratory at Oxford University, and was a member of the team at the NBS that polarized nuclei in 1952. She called Ambler from New York on June 4, 1956, and officially invited him to collaborate on this experiment that would change history.

On the phone, Wu mentioned the paper of Lee and Yang, which suggested several experiments to test their hypothesis. One of them was to use cobalt-60 as the source of beta decay. She knew from the literature that Ambler had been successful in polarizing cobalt-60 several years earlier.

Ambler was not familiar with the beta decay effect in this proposed experiment. He asked Wu if the parity nonconservation effect would be large enough to measure. She was very positive. Ambler became very interested. He asked her to send a copy of the Lee–Yang preprint paper, and was happy to join the experiment.[20]

Being a low-temperature physicist, Ambler did not know of the reputation and achievements of Wu as a nuclear physicist. He called another nuclear physicist, George Temmer, at the NBS. Temmer was an experimental

Ernest Ambler was the principal collaborator of Wu on the experiment studying parity nonconservation.

▼

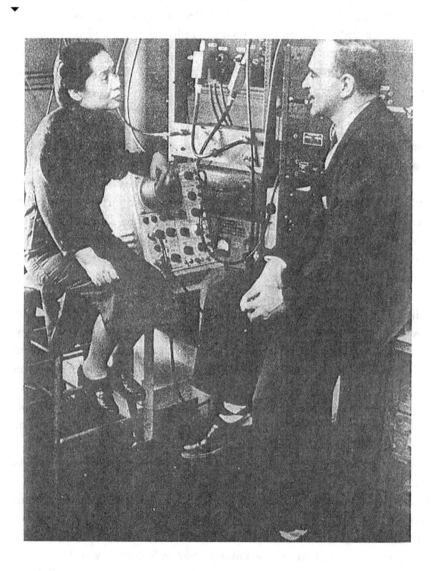

physicist, and had also been a student of Segrè at Berkeley. He had collaborated with Ambler earlier on the nuclear polarization experiment. Temmer had been exiled to the US from Austria as a political refugee, and in the McCarthy era of the late 1950s, his loyalty to the US was questioned. As a result, Temmer was forced to leave the NBS, a government office.

Ambler said on the phone: "George, a physicist, Wu Chien-Shiung of Columbia University, just called. The experiment she proposed is very interesting. How good is she? Shall I join her?" Temmer replied: "She is very tough!"[21]

When Wu was actively preparing for the experiment, the Lee–Yang paper was well circulated. Most of the physicists were skeptical of parity nonconservation. The group led by Telegdi at the University of Chicago was the only other one seriously preparing for an experiment.

From early June to late July, Wu had done numerous tests of the possible effects in nuclear physics at very low temperature, including some that were minute. She said later that if she had known how large, and hard to miss, the effects in the measurement of parity nonconservation would be, she could have saved a lot of the time spent in careful testing. But she still believed that comprehensive preparation was necessary.[22]

Wu wrote Ambler a letter on July 24, telling him that she had successfully completed the preparation for measuring beta decay in a low-temperature environment created by liquid helium. Pending no unforseen technical problem, she suggested a meeting in person, and making a proper arrangement with the administration of the NBS. Ambler replied a week later. His letter discussed low-temperature issues and techniques, both of which helped Wu to formulate better ideas. But Ambler also said that he would not be available to start the experiment as he had planned to take a two-week vacation starting August 4. They apparently had a very different sense of urgency.

Wu had given up her trip back to Asia with Luke for the simple reason that she wanted to do this important experiment as soon as possible, and had hoped to get the result before the physics world realized the significance of this issue. She had been totally devoted to the preparation in June and July. Now, when the experiment was ready to start, her principal collaborator wanted to take a holiday! She was certainly unhappy and anxious. This planted the seeds for a rather unpleasant future collaboration with the scientists at the NBS.

Although the experiment could not start, Wu did not stop her preparation. She used the unexpected "downtime" to carefully examine the possible beta decay effects at low temperature.

In the middle of September, she finally met Ambler in Washington DC. She came away with a nice first impression of Ambler, who later became the Director of the US National Bureau of Standards. Wu recalled that he was soft-spoken, competent, efficient and, most importantly, confident. She had gotton the same impression from their numerous phone conversations before they met in person.

Wu visited Ambler's laboratory, and was introduced to his boss, Ralph Hudson. Ambler and Hudson had both come from the Clarendon Laboratory at Oxford University, under the low-temperature physics authority Nicholas Kurti. They collaborated on many experiments in low-temperature physics at the NBS, including nuclear polarization. Hudson also joined the experiment.

As the experiment would involve detection and measurement of electrons emitted in beta decay, and the measurement of the anisotropy of gamma rays to confirm the polarization of the source (emitter), many electronic instruments were needed. They asked Hayward, to lend them equipment. The two students whom Wu had sent to the NBS did not get along very well with the scientists there. With these developments, Ambler suggested that Hayward and his research student Dale Hoppes be substituted for the students. The official research group was composed of all scientists from the NBS except Wu.[23]

When Wu was actively preparing for the experiment, other laboratories with special traditions, including the ones at Oxford and Rotterdam, also contemplated the possibility. But most of the scientists were not optimistic about being successful in proving parity nonconservation experimentally. None actually did.

The experiment Wu proposed was simple in concept: it would use a strong source of beta decay, properly polarize the source in a certain direction, and then measure the beta rays emitted to determine whether they were tied to the polarization direction. But the experimental design and execution to test this simple concept would prove very difficult and complex.

Cobalt-60 was chosen as the source. It emits tens of thousands of electrons a second (a very good source), and more importantly it changes its spin but not parity after the beta decay. The source had to be polarized. Ambler

132

◄

The three physicists of
the National Bureau of
Standards working with
Wu on the experiment
studying parity
nonconservation:
(*from right to left*) Ernest
Ambler, Raymond W.
Hayward, and Ralph
P. Hudson.

▶

Ralph P. Hudson, Ernest Ambler, Dale D.
Hoppes, and Raymond W. Hayward
(*from left to right*) in front of the
experimental setup.

had previously found that cobalt-60 had to be attached to a crystal, before a strong magnetic field was applied.[24] The source and crystal had to be placed in a very-low-temperature environment to protect the polarization of the nuclei from being disturbed thermally. The extremely low temperature required (0.01 K) was obtained in two steps: using liquid helium to lower the temperature to −270°C, followed by adiabatic demagnetization [by the tiny box made of cerium magnesium nitrate (CMN) crystals] to lower the temperature further to almost −273°C.

Wu's research group at Columbia produced several crystals containing the source. When she placed the crystal in the cryostat in Washington DC, the polarization of the source was found to last only a few seconds, too short to do any measurement. The reason was that the radiation from the nuclei created heat, minute amounts of which raised the temperature, and caused the polarization to disappear. A (nontrivial) solution was to produce a larger crystal that would enclose the smaller crystal source, preventing the heat from leaking out.

Growing crystals is a specialty of chemistry. Wu consulted crystal chemists, and learned that it took long periods of time and precise equipment to grow crystals of the size they had in mind. Wu had neither the funding nor the time. She asked her assistant, H. Fleishman, to search the chemistry department library and locate all references on crystal growth. What he found included a dust-covered, thick manual published in Germany half a century before. Wu learned a lot from this manual, and with the assistance of her students, started to grow crystals in the basement of Pupin. They were able to produce crystals several millimeters in size, which were not large enough.

Wu's research student Marion Biavati brought home the chemical components one evening, and left the glass beaker on the counter near the stove while making dinner. The components were melted by the heat from the stove. The next morning, she found a large beautiful crystal, about 1 cm in size, in the beaker. Wu was delighted by the accidental discovery. She would melt the components with a lamp, and cool them down uniformly, to mass-produce large crystals. They labored for three weeks and produced ten large, perfect crystals.

Wu later recalled that she was the "happiest and proudest person in the world" when she delivered these treasure-crystals to Washington DC.[25]

With these "crystals as beautiful as diamonds" (so said Ambler), Wu and the four scientists from the NBS formally started their experiment.

Their experiment was unusually complex and precise. They encountered many unexpected problems, and proceeded with difficulty.

For example, one had to enclose the smaller crystal source inside a large crystal (to shield it from radiation heat). They learned from a crystal specialist that a dental drill (it exerts its pressure inward and would not shatter the crystals) would work to open a small hole. But the usual sealer (to close the hole after insertion) might not work under extreme low temperature. They tried soap, and even used tiny nylon wires to sew up the hole. They also had to prevent the liquid helium from becoming a superconductor and leaking out; and they used a Lucite rod to transmit the measurement of beta rays. It took a great deal of hard work and the collective experience of Wu and company to overcome the problems.

During this period of time, Wu still had to teach and do research at Columbia University, and she commuted every week between New York and Washington. They were all excited when they observed a huge effect in November. Wu rushed to DC, believing that the effect was too large. They examined the experimental setup and surely enough found components that had collapsed under the action of the magnetic field, causing the false alarm.

They rearranged the setup, and observed a smaller effect in mid-December. Wu was convinced that it was what they were looking for. C. N. Yang believed that such keen insight was the very reason behind her success as a distinguished scientist.

Wu was well known for her caution and precision in doing experiments. After seeing the preliminary result, they wanted to get better confirmation with higher precision before announcing the result publicly. At the same time, Wu asked her research associate to do a theoretical calculation to see if the experimental data were really demonstrating parity nonconservation effect in beta decay.

During this time, Wu met with Yang and Lee at Columbia, and they inquired about the status of the experiment. Wu said that they were quite sure that they had a confirming result, and that M. Morita, a new member of her team and a theoretical physicist from Japan, had completed a calculation. Morita concluded that the beta decay of cobalt-60 was a pure Gamow–Teller transition, and therefore the choice of cobalt-60 as a source had made their experimental result more reliable.[26]

▶

Setup of the experiment studying parity nonconservation that started a scientific revolution.

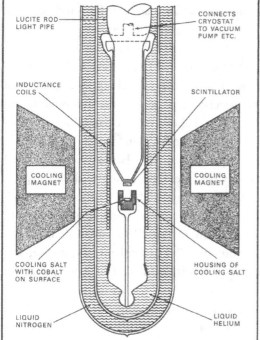

LUCITE ROD LIGHT PIPE

CONNECTS CRYOSTAT TO VACUUM PUMP ETC.

INDUCTANCE COILS

SCINTILLATOR

COOLING MAGNET

COOLING MAGNET

COOLING SALT WITH COBALT ON SURFACE

HOUSING OF COOLING SALT

LIQUID NITROGEN

LIQUID HELIUM

◀

Wu's blueprint of the experimental setup for studying parity nonconservation.

Morita worked with Wu for almost six years, from September 1956 to May 1962. He was the first recipient of the Nishina Research Fellowship, in memory of the "father of nuclear physics in Japan", and made significant contributions to physics research.

As Wu's experiment was progressing, there were more discussions on possible parity nonconservation in the physics world, with different stories and rumors flying around. People familiar with the issue just half a year before must have been surprised to see the sudden change of scenery.

When preparing for his paper with Lee, Yang gave a seminar at MIT in early June. He suggested a careful examination of parity conservation solely in weak interactions as a way of solving the $\theta-\tau$ paradox, and received the immediate attention of Norman Ramsey in the audience. Ramsey would win the Nobel Prize in 1990 (when he was 74 years old) for a technique to measure atomic energy levels that he had developed many years before. He wanted to start an experiment right away to test the Lee–Yang hypothesis.

Ramsey knew that the Oak Ridge National Laboratory had the necessary equipment to achieve low temperatures, and applied to do an experiment there. As another physicist there needed the equipment, and physicists in general were not optimistic about the success of a parity non conservation search, Ramsey did not pursue his plan.[27]

At that time, many scientists visited the Brookhaven National Laboratory in the summer. Ramsey was there, preparing to do his experiment. Richard Feynman, also visiting, stopped by his office and asked, "What are you doing?" Ramsey replied: "I am preparing an experiment to examine parity conservation in weak interactions." Feynman said: "It is a crazy experiment. Don't waste your time." The brilliant, inquisitive, and colorful Feynman even suggested a bet with odds of 10,000 to 1 that the experiment would not succeed.

Ramsey said he still wanted to do the experiment. He reasoned: "If I succeed, my student and I will get a Nobel. If I fail, my student will still get his Ph.D. degree." He and Feynman made a compromise and reduced the odds to 50 to 1. He could not do the experiment at Oak Ridge, and the scientific world was all excited after Wu's result had come out. Parity conservation again became a hot topic at the Seventh Rochester Conference, in April 1957. Feynman saw Ramsey at the conference, and promptly wrote a check for $50 to settle the bet.[28] In 1986, during the "Thirty Years Since Parity Non-conservation" conference at Columbia University, Sidney Drell, the Deputy Director of the Stanford Linear Accelerator Center, told another story. He had attempted to explain the possibility of parity nonconservation in weak interactions to Martin Block at Stanford in the summer of 1956. Block had won the Nobel Prize in 1952 for the discovery of nuclear magnetic resonance. Drell tried very hard, but Block was known to be skeptical and opinionated. In the end, Block said that he would eat his hat if parity nonconservation ever had experimental proof.

Drell said that Block did have a hat. He repeatedly reminded Block to fulfill the promise, even suggesting suitable condiments to go with the hat. Block never paid his bet.[29]

Pauli, also with a sharp tongue, was skeptical about the possibility of parity nonconservation. In 1956, he learned from a letter of his former student Victor Weisskopf that Wu was working on the experiment. He immediately replied that the experiment was a waste of time in his opinion, and that he would bet anything that parity would be conserved.

Weisskopf received the letter just as he learned that Wu was successful. He did not suggest a bet of $1000, but just told Pauli the shocking result. Pauli later wrote to express his astonishment, and joked: "I am glad that I did not conclude our bet. I can afford to lose some of my reputation but not my capital."[30]

There is another Pauli story. George Temmer, a colleague of Ambler and Hudson, was awarded a Guggenheim Fellowship to visit Europe in the second half of 1956. He learned in a letter from Ambler that the experiment had shown some results. He met Pauli at Zurich University, and asked him about the possibility of parity nonconservation in weak interactions. Pauli, in his usual blunt manner, said: "A good experimentalist like Wu should find something more important to do, rather than wasting her time in this obvious matter. Everybody knows that parity is conserved." Several months later, Temmer met Pauli again, at the Bohr Institute in Copenhagen. Pauli did remember his face, if not his name. He said: "Yes, I remember our conversation in Zurich. This matter is finished."[31]

Their experiment was almost done when Wu returned to New York for Christmas in 1956. But finding it hard to believe that nature could be so strange, she was worried that they might have made some mistake.

Although she informed Lee and Yang of the latest result, she asked them not to say anything openly as she wanted to double-check her results.

But Lee did not take it very seriously. On January 4, 1957, he mentioned the new experimental result at the "Friday lunch" of the Columbia physics department. Leon Lederman was present. He realized quickly that his ongoing experiment with the pions and muons, with luck, could be modified slightly to examine the question of parity nonconservation.

Lederman was educated at Columbia University, and stayed there to teach. He left in 1978 to become the Director of the Fermi National Accelerator Laboratory. Before he won the Nobel Prize with two other physicists in 1988 for an experiment they did in the 1960s, he was known for missing great opportunities. As an example, when Samuel C. C. Ting and Burton Richter discovered a new particle in 1974, Lederman was doing a very similar experiment searching for new particles. The energy spectrum he found did not show a sharp peak (as did the one in Ting's experiment indicating a particle) but instead showed a smooth plateau like a shoulder. It was jokingly called "Lederman's shoulder".

Lederman originally thought that even if parity was not conserved, the effect would be very small. When he learned that Wu's result was surprisingly large, he decided to check it out. He called his colleague Garwin. They agreed to meet, following their phone discussion that night, in thirty minutes at the Nevis Accelerator Laboratory, north of Columbia University.

M. Weinrich, a graduate student of Lederman, was at Nevis working on the pion experiment. Like other graduate students, Weinrich was doing the boring work of recording data or checking equipment. Lederman and Garwin suddenly showed up, pushed him aside, took over, and rearranged the equipment.[32]

By 10 o'clock, Lederman and Garwin figured out how to proceed. The experiment took only four days, and showed rather definitive results. Garwin had made a critical contribution to this clever design.

At six o'clock in the morning of January 8, Lederman called Lee to announce: "The principle of parity conservation is dead."

Wu returned to the NBS on January 2. She and her four collaborators carefully examined their experiment. She recalled that their work was most intense from January 2 to 8, when they repeatedly lowered the temperature of the cryostat to search for any factors that would invalidate their result.

The graduate student Hoppes would sleep in a sleeping bag on the floor, and call them whenever the temperature was low enough so that they could rush back to the laboratory in the cold nights.

By January 7, the news of the successful experiment of Lederman and Garwin (pion → muon → electron + neutrino) was widely circulated. The high level officials at the NBS were anxious to find out the status of Wu's experiment.

Wu was under tremendous pressure from the news that another experiment had proven parity nonconservation in weak interactions. She did not panic, and continued with the various checks. At two in the morning of January 9, they finally finished all scheduled tests. All five of them celebrated this great moment in the laboratory. Hudson smiled and took out a bottle of French red wine, vintage 1949, and paper cups. They drank to the overthrow of the principle of parity conservation.[33]

The next morning, other low-temperature physicists were surprised to find their laboratory relaxed and quiet. They saw the empty bottle and paper cups in the garbage can, and realized: "Great, parity conservation in beta decay is dead!"[34]

Scientists always like to publish their research results in a scientific journal as soon as possible. The editors of a journal will send the paper out for peer review before they decide whether to put it in print or not. In any case, the journal will always note the "date received" as a time stamp. The "date published" and "date received" will become important time indicators, and determine who will get the credit for the scientific results.

After their experiment was done, Wu and her collaborators obviously wanted to write a paper. As Wu originated the experiment, and had the best understanding, she wrote a report near the end of the experiment in private. The four scientists of the NBS and Wu sat down the following Sunday to discuss the writing of the paper. Wu pulled out a completed paper. The four scientists were surprised that Wu would write it up without any discussion with them. They originally thought that the experiment was a collaboration of equals, and now realized that Wu had regarded them as helpers in her experiment all along.[35]

In addition, the four scientists were unhappy with the writing: the paper referred only to the Lee–Yang paper and their discussions with Wu, with no mention of them. No one could change the writing.

Then there was the listing of the authors. An alphabetical order would list Ambler first and Wu last. With a sigh, Wu said that this would not be the correct approach. So, like the perfect Englishman, Ambler asked: "Would you like to go first, Miss Wu?" The four at the NBS in alphabetical order of names followed.[36]

The paper almost marked the end of Wu's collaboration with the scientists at the NBS, as their relationship deteriorated rapidly. There were cultural conflicts. Wu was anxious to start the experiment, but Ambler had to go on vacation for two weeks. During the experiment, the easy-going mood of the scientists at the NBS was very different from her style. She believed that lunch should last no longer than 15 minutes, and found it unthinkable for the members to continue playing bridge after lunch. The scientists at the NBS felt that Wu was very insecure.[37]

Wu went to Columbia University in 1944 as a senior scientist, became an associate, and then an associate professor in 1952. She had done first-rate research on beta decay. Her many colleagues at Columbia agreed that she was not treated properly there, partly because she was a woman scientist, and partly because of her status as a naturalized citizen.[38] Given the conservative mentality of the Ivy League universities in the East, Columbia University employing a woman professor was indeed ahead of its time.

Wu had significant achievements from 1952 to 1956, but was not further promoted. One could feel the pressure she was under. The Columbia physics department was a heavyweight institution in the 1950s, with many top scientists gathering there. Wu had the urge to establish her position there, and to do a great experiment testing parity conservation.

In the past, she had always done experiments in universities, mostly with her students or colleagues, and she was well known for her tough demands on the students. She obviously could not treat the scientists at the NBS the same way. Hudson, now with the National Science Foundation, recalled that they did not readily take orders from Wu, and as a result she was not that happy in those several months. When she was unhappy, she would keep very quiet, and sigh mournfully to other team members.

Ambler said that Wu was probably not used to working with two British men and two American men. She spoke conservatively, sometimes even ambiguously, which contributed to poor communication. Hudson said that Wu was quite depressed during that period.[39]

There was no question that the scientists at the NBS respected her authority on beta decay. Her status and accomplishments were at a higher level. They also agreed that they would not, and could not, perform the experiment without her initiation and expertise. They did resent the fact that they were referred to as merely technicians, but Wu was the star of the experiment, receiving all the glory and spotlight.[40]

The paper by Wu and company arrived at the *Physical Review* on January 15, 1957. The Lederman paper, reporting on the experiment that had been finished just a couple of days before that date, arrived that same day. It also stated the fact that the experiment was initiated only after they learned about the definitive result of Wu's experiment. Both papers were published in the February 15 issue of the *Physical Review*.

Otto Frisch of Cambridge University, who had made major contributions through nuclear fission experiments, said in a lecture at that time: "The statement 'parity is not conserved' is difficult to comprehend, but spreads to the world as a new gospel."[41] People believed that Wu's experiment was as important as the one determining the speed of light by the American scientists A. Michelson and E. Morley in 1887. The Michelson–Morley experiment blazed the trail for the theory of relativity by Albert Einstein.

To the public, parity nonconservation is still rather ambiguous. Some mistakenly think that it has overthrown the theory of relativity. To date, parity nonconservation has not had any practical value.

To scientists, parity nonconservation is a revolutionary development. Wu could not sleep for two weeks after the experiment was done. She kept asking herself why God had picked her to reveal this secret. She said: "We learn one lesson: never accept any 'self-evident' principle."[42]

On January 15, the day when both the Wu paper and the Lederman paper arrived at *Physical Review*, Columbia University held an unprecedented press conference announcing the discovery. Rabi, the most senior physicist at Columbia, was on sabbatical leave at MIT. He was called back to host the press conference. This was held in the Pupin Laboratories at two in the afternoon. All the Columbia scientists involved in the discovery were there, including Wu, Lee, Lederman, and Garwin.

C.N. Yang of the Institute for Advanced Study was not present. In *Selected Papers (1945–1980) with Commentary* (published in 1983), celebrating his 60th birthday, Yang wrote: "I believe it is not appropriate to

announce a scientific development in a press conference."[43] The scientists
at the NBS were also uneasy about this kind of public relations event.[44]

The world-renowned *The New York Times* reported this scientific event
on the front page the following day. The headline was "Basic Concept in
Physics Is Reported Upset in Tests". This rather long report included the
announcement at the press conference, the materials released by Columbia
University on the challenge of parity conservation by the theorists and
experimentalists, and the meaning of parity conservation. It quoted Rabi: "In
a certain sense, a rather complete theoretical structure has been shattered at
the base, and we are not sure how the pieces will be put together."[45]

On January 17, *The New York Times* published an editorial with the
title "Appearance and Reality". Near the end, it wrote: "This, it is believed,
has removed the principal roadblock against the building of a comprehensive theory about the fundamental building blocks of which the material
universe is constituted. What the theory will be may take another 20 years
in the making, but physicists now feel confident that they have at last found
a way out of the present 'cosmic jungle'."[46]

Important magazines, such as *Time* and *Life*, also reported this big
event extensively in that month. In these reports, Lee and Yang, who proposed the theoretical framework, were naturally central figures, and Wu,
who provided the first experimental evidence for their hypothesis, was
highly praised.

After the news was announced, physicists rushed to repeat the experiment in their laboratories. Wu received many inquiries, as well as
congratulations. Her old friend Pauli wrote her a letter on January 19. He
said that he was still puzzled by the fact that parity conservation did not hold
in weak interactions but held in strong interactions; he would still congratu-
late her on her success. The puzzle does not have an answer as of today.

Very few revolutionary events are not accompanied by pain and difficulty. For Wu and the scientists at the NBS, the experiment itself, and the
collaboration process were very difficult. Pain was felt by another experimentalist, Valentine Telegdi, who also independently proved parity
nonconservation.

Telegdi was a distinguished Hungarian physicist (other Hungarian
physicists included Eugene Wigner, Edward Teller, and Theodore von
Kármán). He examined parity conservation in weak interactions about the

same time as Wu. The experiment was delayed when he went to Europe for his father's funeral. He returned and learned of the result of Wu's experiment, and rushed to finish the experiment and the paper.

The paper was rejected at first, and later published two weeks after the papers of Wu and of Lederman, in the next issue of the *Physical Review*. It had the same "date received". Telegdi was furious, believing that the editors were unduly influenced by "nonscientific factors". He even suspected that the "Columbia University gang" had played a trick on him.[47]

S. Goulsmit, the Editor-in-Chief of the *Physical Review*, later explained that Telegdi's paper was delayed because it did not have enough data.[48] Telegdi was still unhappy. He announced his withdrawal from the American Physical Society (publisher of the *Physical Review*) and even considered leaving physics research altogether.[49] Garwin, a high-level executive at IBM, promised him: "Anytime you want to come to IBM, you have a job here."[50] Telegdi did not leave physics, and made important contributions afterward. He won the Wolf Prize (the so-called Israeli Nobel) in 1990.

The annual meeting of the American Physical Society was held in the New Yorker Hotel on January 30 and had a record attendance of more than 3000, all because of the excitement over the discovery of parity nonconservation. In the special session on the last day, Wu reported on her experiment. Yang, Lederman, and Telegdi also reported. The groups led by R. R. Rau of the Brookhaven National Laboratory, and by Luis Alvarez of UC Berkeley reported on their similar experimental results. There was a parallel session in astrophysics, but everyone stayed in the jam-packed ballroom.[51]

As this experimental development had much impact, Wu received numerous invitations to speak at universities and research institutes. Although she never liked to travel, she was on the road a lot.

Wu was invited to speak at the Seventh Rochester Conference in April. The Rochester Conference was a particle physics meeting dominated by strong interaction issues. Wu said during the lecture: "I am not working in your field. I'm here on the strength of the weak interaction."[52] Her travel and lectures were not limited to the US, but included Switzerland, France, Italy and Israel, and she received attention everywhere she went.

In October 1957, Wu was lecturing at a university in upstate New York when a call came, announcing the caller as Oppenheimer. He told the operator

144

In 1957, C. N. Yang and T. D. Lee (*front row, leftmost two*) were awarded the Nobel Prize in Physics.

▼

r. Chien-Shiung Wu

r. Chen Ning Yang

A Historic Colloquium
in Palmer Laboratory

At a 1957 conference called by Professor John A. Wheeler
Dr. Chien-Shiung Wu (above) and Dr. Chen Ning Yang
(left) explain to an awe-struck audience their revolutionary theory and ingenious experiment which together
shattered one of the most deeply entrenched principles of
physics—the principle of parity or invariance (that nature, loving symmetry, makes no fundamental distinction
between right and left)—first inserted into quantum mechanics in 1927 by Professor Eugene P. Wigner. At the
conclusion Miss Wu asked for questions: the response
was dead silence for two minutes, then thunderous applause and a standing ovation. For this Dr. Yang and his
colleague, Dr. Tsung Dao Lee, now both at the Institute
for Advanced Study, received the Nobel Prize (and with
Miss Wu, Princeton honorary degrees).

Front row (r. to l.): Dr. R. L. Garwin, Dr.
Robert Oppenheimer, Dean Donald R. Hamilton '35, Dr. T. H. Berlin, Dr. Yang, Professor George T. Reynolds

At a colloquium in 1957, Wu together with C. N. Yang explained their revolutionary experiment and theory of parity nonconservation, to a full room of physicists including J. Robert
Oppenheimer (*front row, second from right, with hand under chin*).

that he would wait by the phone for her to finish her lecture. He simply told Wu: "Gee Gee, Yang and Lee won the Nobel Prize for this year."[53] Oppenheimer was the Director of the Institute for Advanced Study in Princeton. He gave a celebration banquet, inviting Wu, Lee, Yang and others.

Everyone gathered in the library before the banquet. Oppenheimer gave a short speech emphasizing that there were three people deserving most of the credit for proving parity nonconservation: Wu, in addition to Lee and Yang. He said that we should not overlook Wu's contribution. He arranged for Wu to sit beside him at the banquet, demonstrating his admiration.[54]

People were surprised and somewhat resentful that the Royal Swedish Academy did not award the Nobel to Wu that year. Many great scientists publicly expressed their disappointment and disapproval. For example, the 1988 Nobel laureate Jack Steinberger felt that it was the biggest mistake that the Nobel Prize was not awarded to Wu also. Lee and Yang conceived of the nonconservation of parity, but it was Wu who confirmed it.[55]

After Steinberger was awarded the Nobel in 1988, the magazine *Science* published an article on the winning experiment of Steinberger, Lederman and Melvin Schwartz (who later founded a computer company) at Columbia. The article mentioned that a total of six Nobel Prizes were awarded to scientists for research done at Columbia University in the 15 years from late 1940 to early 1960.[56] A footnote specially stated that Steinberger and others thought that Wu should be a co-recipient of the 1957 Nobel Prize.[57]

Wu had yet to receive a Nobel. The deliberation of the selection committee of Nobel Prizes is under seal for 50 years. When it becomes public in 2006, one could have a better idea of how the decision was made. In the meantime, the following may offer an explanation.

According to the regulations for the Nobel Prize, there cannot be more than three recipients in a field in the same year. Wu, Lee, and Yang would make three. Would not her collaborators — especially Ambler — also be included?

Ambler was educated in the low-temperature laboratory at Oxford University. He made a definitive contribution to the experiment. As Wu was much better known in physics, people referred to the experiment as Wu's experiment. She herself frequently referred to it as one done by Wu *et al.* in her lectures. The scientists at the NBS believed that it was a joint effort. Without their technology of nuclear polarization at low temperature, the experiment simply could not have been done.[58]

They believed that both the pressure at Columbia University and the desire to achieve at all costs made her unwilling to share credit with collaborators.[59] The fact that scientists tended to give her all the credit could be glimpsed through the following story.

George Temmer, a colleague of Ambler, was in a conference at Harvard shortly after the parity conservation experiment had been completed. Someone said: "See, the experiment was just done." Temmer asked: "Which experiment?" The same person said: "The parity conservation experiment done at *Columbia University*." Temmer said: "Columbia? It was done at the *National Bureau of Standards*." "All right, the National Bureau of Standards?" Temmer insisted: "You are all wrong. I was there. They told me that they were working on the experiment."[60]

Temmer believed that this general accounting was unfair. He later wrote a magazine article arguing that the experiment could not have succeeded without the technology at the NBS. The scientists at Oxford were also angry about this lopsided accounting. Nicholas Kurti, a pioneer in low-temperature physics and a professor of both Ambler and Hudson, was unhappy that his students were not getting the credit they deserved.

Kurti published an article in *Physics Today* in March 1958. The title was "Nuclear Orientation and Nuclear Cooling". It discussed the developmental history and the importance of these techniques, and especially pointed out that the critical role of low-temperature physicists in the parity conservation experiment was sadly overlooked.[61]

With this argument in the background, the opinion in England might have influenced the decision in Sweden. Many Nobel laureates said that they had nominated Wu for the Nobel Prize, so she was a shortlisted candidate many times.

She did not want to reveal her own feelings publicly about not receiving the Nobel Prize. In a letter she wrote to Steinberger in January 1988, she congratulated him on the 1988 Prize, and expressed her genuine appreciation of his praise in *Science*. She wrote: "I treasure and value your praise, coming from a modern-day physicist with such a critical mind, more than any prize and honor in science. I dedicated my whole life to, and found happiness in, research in weak interactions. Although I did not do research

In 1958, Wu received the Research Corporation Award;
with the President of Columbia University, G. L. Kirk
(*left*), and Nobel-winning physicist C. N. Yang (*right*).

just for the prize, it still hurts me a lot that my work was overlooked for certain reasons."[62]

The revolutionary parity nonconservation theory was regarded as the most important achievement in physics in that decade. Emilio Segrè, a professor of Wu, even said in his book that it was possibly the greatest discovery after the Second World War. Wu was a critical figure in this discovery. When interviewed by the *New York Post* in 1962 after receiving the "Woman of the Year" award from the American Association of University Women, she said: "There has been much progress in research of parity nonconservation since 1956, but nobody knows where it will lead us.... In 1906, the young Einstein had no idea whether his formula $E = mc^2$ would be useful. The answer came, after thirty five years, when the first prototype reactor was built under the athletic field at the University of Chicago."

The overthrow of the principle of parity conservation has special meaning in the Chinese scientific culture, as the three key figures were all Chinese.

Segrè wrote in his book: "The achievements of these three Chinese physicists indicated that if China can overcome its recent political chaos, and resume its historical leadership position of world culture, its contribution to physics will be marvelous."[63]

Notes

1. "One Researcher's Personal Account", C. S. Wu, in *Adventures in Experimental Physics*, ed. B. Maglich, p. 116.
2. *Nobel Foundation Directory, 1989–1990*, and "This Was the Particle Physics", G. Feinberg, in *Celebration of the 1988 Nobel Prize in Physics*, February 17, 1989.
3. Interview with Luke Yuan, September 10, 1989, New York residence.
4. "Present Knowledge About the New Particles", C. N. Yang, in *Selected Papers (1945– 1980) with Commentary*, p. 194. Lecture delivered in Seattle, September 1956.
5. *High Energy Nuclear Physics: Proceedings of the Sixth Annual Rochester Conference, April 3–7, 1956*, ed. J. Ballam *et al.*, Interscience, New York, 1956.
6. "Commentary of 'Introductory Talk' at the 1956 Rochester Conference, Session on Theoretical Interpretation of New Particles", in *Selected Papers (1945–1980) with Commentary*, C. N. Yang, p. 25.
7. "Commentary of 'Question of Parity Conservation in Weak Interactions' ", in *Selected Papers (1945–1980) with Commentary*, C. N. Yang, p. 26. W. H. Freeman & Company,

1982. "*Broken Parity*", T. D. Lee, in *T. D. Lee: Selected Papers*, ed. G. Feinberg, p. 487, Birkhauser, Boston, 1986. Lee and Yang had very different accounts of how the concept of parity nonconservation was created and by whom. This was the major factor causing their breakup.

8. *A Question of Parity*, Jeremy Bernstein, New Yorker, May 12, 1962.

9. Interview with Wu Chien-Shiung. "One Researcher's Personal Account", C. S. Wu, in *Adventures in Experimental Physics*, ed. B. Maglich, p. 102. "Reminiscences", T. D. Lee, in *Thirty Years since Parity Non-conservation*, ed. R. Novick, pp. 162–163, Birkhauser, Boston, 1988.

10. "One Researcher's Personal Account", C. S. Wu, in *Adventures in Experimental Physics*, ed. B. Maglich, p. 104.

11. Interview with Maurice Goldhaber, April 10, 1990, Brookhaven National Laboratory.

12. *International Colloquium on the History of Particle Physics: Some Discoveries, Concepts, Institutions from the Thirties to Fifties*, July 21–23, 1982, C8–450.

13. Interview with C. N. Yang, November 16, 1989, New York City.

14. V. L. Telegdi was the leader of another research group testing parity nonconservation at the University of Chicago.

15. *International Colloquium on the History of Particle Physics: Some Discoveries, Concepts, Institutions from the Thirties to Fifties*, July 21–23, 1982, C8–450.

16. "One Researcher's Personal Account," C. S. Wu, in *Adventures in Experimental Physics*, ed. B. Maglich, p. 104.

17. Interview with Luke Yuan, March 4, 1990, New York residence.

18. Richard Garvin in *Current Biography*, ed. C. Maritz *et al.*, Vol. 50, No. 3, March 1989, H. W. Wilson Company.

19. Private letter to the author from R. Garwin, February 23, 1990.

20. Interview with Ernest Ambler, December 8, 1989, Washington DC.

21. Interview with Ernest Ambler, December 8, 1989, Washington DC.

22. "One Researcher's Personal Account," C. S. Wu, in *Adventures in Experimental Physics*, ed. B. Maglich, p. 109.

23. Interview with Ernest Ambler, December 8, 1989, Washington DC.
    Interview with Hayward, March 22, 1990, National Bureau of Standards, Gaithersburg, Maryland. Interview with R. P. Hudson, March 23, 1990, National Bureau of Standards. "One Researcher's Personal Account", C. S. Wu, in *Adventures in Experimental Physics*, ed. B. Maglich, pp. 110–111.

24. The crystal was cerium magnesium nitrate (CMN).

25. "One Researcher's Personal Account", C. S. Wu, in *Adventures in Experimental Physics*, ed. B. Maglich, p. 115.

26. Same as 25.

27. Interview with Norman Ramsey, February 22, 1990, Harvard University.

28. Interview with Norman Ramsey, February 22, 1990, Harvard University.

29. *Thirty Years Since Parity Non-conservation — A Symposium for T. D. Lee*, ed. R. Novick, p. 86, Birkhauser, Boston, 1988.

30. *The Privilege of Being a Physicist*, Victor F. Weisskopf, pp. 164–165, W. H. Freeman and Company, 1989.

31. Interview with George Temmer, December 12, 1989, Rutgers University.

32. Interview with Leon Lederman, March 27, 1990, office at University of Chicago. Interview with C. N. Yang, September 12, 1989, SUNY Stony Brook.

33. Wu did not drink, as she had never been a drinker.

34. "One Researcher's Personal Account", C. S. Wu, in *Adventures in Experimental Physics*, ed. B. Maglich, pp. 117–118.

35. Interview with R. P. Hudson, March 23, 1990, National Bureau of Standards.

36. Interview with Hayward, March 22, 1990, National Bureau of Standards, Gaithersburg, Maryland.
Interview with R. P. Hudson, March 22, 1990, National Bureau of Standards. Interview with R. P. Hudson, March 23, 1990, National Bureau of Standards.

37. Interview with Hayward, March 22, 1990, National Bureau of Standards, Gaithersburg, Maryland.
Interview with R. P. Hudson, March 23, 1990, National Bureau of Standards.

38. The colleagues of Wu at Columbia University, Jack Steinberger, T. D. Lee, Robert Serber, Leon Lederman, and Gerald Feinberg all thought the same.

39. Interview with Ernest Ambler, December 8, 1989, Washington, DC.
Interview with R. P. Hudson, March 23, 1990, National Bureau of Standards.

40. Interviews with Hudson, Hayward, and Ambler.

41. "One Researcher's Personal Account", C. S. Wu, in *Adventures in Experimental Physics*, ed. B. Maglich, p. 118.

42. *Particular Passions — Talks with Women Who Have Shaped Our Times*, Lynn Gilbert and Gayle Moore, p. 70, Clark N. Potter, Inc., 1981.

43. "Commentary of 'Parity Non-Conservation and a Two-Component Theory of the Neutrino'," in *Selected Papers (1945–1980) with Commentary*, C. N. Yang, p. 36, W. H. Freeman & Company, 1982.

44. Interview with R. P. Hudson, March 23, 1990, National Bureau of Standards.

45. *New York Times*, 24th edition, front page, January 16, 1957.

46. *New York Times*, editorial, January 17, 1957.

47. Interview with Valentine Telegdi, May 22, 1990, CERN, Geneva.

48. "A Reply from the Editor of the Physical Review", S. Goudsmit, in *Adventure in Experimental Physics*, ed. B. Maglich, p. 137.

49. "One Researcher's Personal Account", V. Telegdi, in *Adventures in Experimental Physics*, ed. B. Maglich, p. 134.

50. "An Introductory Remark", V. Teleghi, in *Thirty Years Since Parity Non-conservation*, ed. R. Novick, p. 38, Birkhauser, Boston, 1988.

51. *Selected Papers (1945–1980) with Commentary*, C. N. Yang, p. 37, W. H. Freeman & Company, 1982.

52. "The Weak Interactions", S. B. Treiman, in *Scientific American*, p. 80, March 1958.

53. Interview with Wu Chien-Shiung, January 6, 1990, New York City residence.

54. Interview with Wu Chien-Shiung, January 6, 1990, New York City residence.

55. Interview with Jack Steinberger, August 9, 1989, CERN, Geneva.

56. In addition to the Nobel Prize awarded to Steinberger, Lederman and Schwartz for their experiment in 1962, the other five Nobel Prizes were awarded to Willis Lamb in 1955 for his precise measurement of the structure of the helium atom, to Polykarp Kusch in 1955 for his measurement of the electron magnetic moment, to T. D. Lee in 1957 for his proposal of parity nonconservation in weak interactions, to Charles Townes in 1964 for his discovery of the laser, and to James Rainwater in 1975 for his research on intranuclear movement.

57. *Science*, p. 669, November 6, 1988.

58. Interviews with Ambler, Hudson, and Hayward, December 8, 1989 and March 22 and 23, 1990.

59. Interview with Hayward, March 22, 1990, National Bureau of Standards.

60. Interview with R. P. Hudson, March 23, 1990, National Bureau of Standards.

61. "Nuclear Orientation and Nuclear Cooling", Nicholas Kurti, in *Physics Today*, p. 22, March 1958.

62. Letter from Wu Chien-Shiung to Steinberger, January 1, 1989, provided by Jack Steinberger.

63. *From X-Rays to Quarks*, Emilio Segrè, p. 260, W. H. Freeman and Company, 1980.

Chapter 10

Experiment at 2,000 Feet Underground

One day in the 1960s, a middle-aged Chinese lady emerged from a salt mine 2000 feet below Lake Erie, near Cleveland, Ohio. She smiled broadly, adjusting to the bright sunlight. She was the distinguished experimental physicist Wu Chien-Shiung.[1]

Wu led an experiment which was conducted in a small room deep underground. The room was kept very dry to protect the experimentalists and apparatus from the acid vapor in the cave; and the thick, dense salt rock above the cave provided an excellent shield against cosmic rays continuously bombarding the earth. The object of the investigation, *double beta decay*, was very rare, and therefore very sensitive to the minute background of cosmic rays.

Investigation into rare double beta decay was a natural extension of Wu's long-time research on beta decay.

The experiment on parity nonconservation that Wu finished in early 1957 had an unprecedented impact on physics. One of the theorists who proposed the possibility, C. N. Yang, recalled forty years later that the experiment was the one that shocked physicists psychologically the most in the half-century since the Second World War.[2]

Wu's experiment confirmed that *parity* was not conserved in weak interactions. It implied that *charge conjugation* was not conserved in weak interactions either.

What is charge conjugation? Simply speaking, it defines the relationship between a particle and its antiparticle. In 1920, the British physicist Paul Dirac first theoretically predicted antiparticles, with identical mass but opposite charge to the corresponding particles. An American, Carl Anderson, discovered the positron in 1932. In physics, charge conjugation invariance means that a natural phenomenon does not change if particles are replaced by anti-particles.

In addition, Wu's experiment revived a "two-component theory of the neutrino". The mathematician H. Weyl first proposed the theory in 1929. As it did not satisfy parity conservation, Weyl and other physicists gave it up.

When parity conservation was questioned in 1956, physicists naturally revived the two-component theory of the neutrino. These physicists included Lee and Yang, the Pakistani Abdus Salam, and the top Russian theoretical physicist Lev Landau. Landau won the Nobel Prize in Physics in 1962 for his theories in condensed matter, especially superconductivity. Salam was one of three winners in 1979.

As Wu's experiment confirmed parity nonconservation, it revived the two-component theory of the neutrino. The theory also implied that the neutrino has zero rest mass. Wu's experimental results on the beta spectrum, especially near the upper energy limit, put an upper limit on the neutrino mass at ~ 200 eV, which is close to zero.[3]

Wu later recalled that their experiments "killed three birds with a single stone": parity nonconservation, charge conjugation nonconservation, and the two-component theory of the neutrino.[4]

As her experiment confirming parity nonconservation had such a lasting impact, she was invited to lecture at many places in the US, and to conduct a long lecture tour in Europe in the second half of 1957. She visited Italy, Israel, Switzerland, France, and England, getting the spotlight and press coverage everywhere. In England, she met an old friend from her National Central University days, the artist Zhang Xi-Ying, and her husband, Fei Cheng-Wu; Lu Gui-Zhen of Columbia University, then working with the historian Joseph Needham at Cambridge University; and Ursula Schaefer from her UC Berkeley days and her husband, Willis Lamb, who had won the Nobel Prize two years earlier and was visiting Oxford.

Israel impressed Wu the most, when she visited it in September. She attended the Rehovot Conference on Nuclear Structure. She wrote to a friend: "The tiny Israel impressed me the most. People there are industrious and great achievers. They have reclaimed two-thirds of the desert and transformed it into agricultural land for settlement."[5]

Rehovot was a scenic resort overlooking the Mediterranean Sea. The scientists stayed in Zion Hotel, reserved for the meeting. The then Prime Minister of Israel, David Ben-Gurion, and his wife also stayed there, with the scientists. One afternoon, Ben-Gurion wanted to discuss with Wu her experiment. Not finding her, he left a note.

Ben-Gurion said in the note that he had recently read an Indian book on yoga discussing the origin of the positive and negative poles, and he was curious to find out if the ideas were consistent with modern physics. When meeting her later, he also asked if the ancient historical ideas were relevant to modern experimental findings. Ben-Gurion's thinking on philosophy and science left Wu impressed.[6]

Wu flew with Pauli from Italy to Israel. Pauli belonged to an earlier generation of physicists, and he still found the proof of parity nonconservation in 1957 puzzling and unbelievable. He wrote Wu almost every day after she had finished her experiment, discussing with her all his questions.[7]

Pauli was working in Switzerland, and was not that familiar with her ongoing experiments after the big one on parity nonconservation. Wu explained her recent results during their flight from Rome to Israel. A little later, Pauli wrote a vivid letter about this journey to his sister, who forwarded it to Wu after he died of illness the following year.

In this letter written in German in September 1957, Pauli described his excitement of discussing the latest results on parity nonconservation. He and Wu apparently had a night flight; he added: "It was memorable traveling with physics and a Chinese immigrant. *Frau* Wu is as obsessed with physics as I was in my youth. I doubt whether she ever noticed the light of the full moon outside."[8]

For Pauli, he assumed that the discovery of parity nonconservation would be as revolutionary as the development of quantum mechanics in the 1920s. Parity nonconservation may be revolutionary, but is not in the same vein as quantum mechanics.

In the early days of the development of quantum mechanics, many theories were rather speculative. Pauli and the German W. Heisenberg had

Mutual admiration in sciences between Wu and Wolfgang Pauli.

their share of speculations, but were very successful. After the discovery of parity nonconservation in 1957, they were excited, thinking that it would be the 1920s all over again. Pauli and Heisenberg investigated a field theory with high-dimensional symmetry called the "universal formula".

When Pauli visited the US in 1958, he asked Wu to invite a few people for a casual discussion. But almost four hundred people showed up at Columbia University and filled the auditorium. Pauli grew more and more uneasy as the lecture on the "universal formula" went on. Many, including Yang, raised questions, and felt that the "universal formula" was built on sand.

Many years later, the physicist F. Dyson of the Institute for Advanced Study in Princeton recalled that he joked, right after the lecture, to another theorist, A. Wightman: "If they were always fooling around like today, maybe we should go back and investigate if their work in 1925 (quantum mechanics) was incorrect."[9]

After the lecture, five of them — Wu, Luke Yuan, Yang, Lee, and Pauli — drove to a Chinese restaurant called Shun Lee Garden for dinner. With Lee driving, Pauli, being short and heavy, was in the passenger seat.

He liked to rock back and forth, and said: "The more I talk about this 'universal formula', the less I believe in it." Yang said that he was depressed hearing that, and afraid that Pauli might start to lose his clear mind in his old age.

But Heisenberg believed in the theory wholeheartedly. He even announced in a German newspaper that he and Pauli had solved an important, fundamental problem in physics. Pauli was not happy, regarding the newspaper article as just an advertisement, and wrote Wu about this incident.

In newspaper articles, first in Germany and then in the US, Heisenberg claimed that their fundamental theory had been completed, except for some details. Pauli, with a sharp tongue, wrote a letter to Heisenberg saying: "I totally disagree with what you said yesterday." He drew a square, with nothing inside. "I could be as good an artist as Titian [a famous painter], but I have not yet put in the details."

At an international high-energy physics conference in Geneva in the summer of 1958, Heisenberg presented the theory that he and Pauli had worked out. Pauli attacked him right afterward. Yang said that he had never seen two prominent physicists attacking each other so relentlessly. Heisenberg was calm, but Pauli was furious. Actually, Pauli was already very sick. He died three months later.[10]

When Wu returned to Columbia University after the European trip, her students told her that a physicist, after spending a summer in Argentina, visited the laboratory right after she left. He inquired about the recent development in parity nonconservation, and the students told him their latest definitive results on the "two-component theory of the neutrino" and "conservation of leptons". The students asked who this physicist was, and were astonished to learn that he was none other than the genius Richard Feynman.[11] (*Author's note*: There might have been a mistake in her memory, as Feynman had spent the summer in Brazil, not Argentina.)

The happy coincidence is that the next very important achievement of Wu is related to this physicist, Feynman.

As a result of Wu's parity nonconservation experiment and other experiments on beta decay, the exact nature of these weak interactions became important. Based on many experimental results, as well as a

Wu in the office of Columbia University.

correction of mistakes in the experiment of Wu's students (see Chapter 8), the *V–A theory* (universal Fermi interaction) was firmly established.

Feynman and Gell-Mann of Caltech wrote a paper in 1957, submitted it in September, and had it published on January 1, 1958. This paper, entitled "Theory of the Fermi Interaction", noted the similarity between ordinary beta decay and muonic decay (there are three major phenomena in weak interactions: beta decay, muonic decay, and muon capture). Based on the new result of parity nonconservation, which implied two-component neutrinos, and conservation of leptons, they proposed a "universal' form of the Fermi interaction, $g*(V–A)$ for *both* beta decay (n → p + electron + anti-neutrino) and muonic decay (electron → muon + antineutrino + neutrino). The value of g in muonic decay differs no more than 3% from the value derived from neutron decay.

Their hypothesis was called *conserved vector current*, an especially elegant formulation of the beta decay.

This hypothesis naturally explained the ordinary beta decay n → p + electron + antineutrino, and n → p + muon + anti-(muon)neutrino. The muon is some 200 times heavier than the electron, both being "leptons".

The hypothesis of conserved vector current was analogous to quantum electrodynamics describing electrons. Conservation of electromagnetic current implies that the total charge (the coupling constant "e") is unaffected by the presence of strong interactions. Analogously, the idea of conserved vector current implies that the total "weak vector charge" (the coupling constant g) is unaffected by the strong interactions.

There was another coincidence, as Feynman was one of the creators of quantum electrodynamics. He won the 1965 Nobel Prize in Physics with the other creators, the American Julian Schwinger and the Japanese Sin-Itiro Tomonaga.

Gell-Mann wrote another paper in the *Physical Review*, in which he proposed examining the beta ray spectra in the decays of radioactive boron (B-12) and nitrogen (N-12) to carbon (C-12) to test the hypothesis of conserved vector current. The CVC theory was very elegant and the proposed experiment was very appropriate, but the experiment was extremely difficult. The theoretical prediction of the beta ray spectra effect of CVC is very minute, easily leading to mistakes if one is not careful. Experiments at UC Berkeley, the Dubna Institute in Russia, and CERN in Switzerland in 1958 all failed.

Gell-Mann knew that Wu was the top physicist in the field of beta decay. At a physics meeting at MIT in 1959, he invited Wu to do the experiment to test their hypothesis, and famously said: "How long did Yang and Lee pursue you to follow up their work?"[12]

Wu did not start right away, as she was really busy then. In addition to tens of graduate students and researchers doing different experiments, she was frequently invited to conferences. She would rather stay in the laboratory, but had to be on the road in that period.

Wu, Luke, and Vincent, then not yet 12 years old, had a vacation in June 1958. They sailed on the French ocean liner *Liberté* across the Atlantic Ocean, and switched to smaller riverboats to visit Switzerland, France, Belgium, and Germany. They stayed in Geneva in August, with Luke doing research and Wu writing papers at CERN, and Vincent learning French. Near the end of September, Wu returned to Columbia University alone, Luke transferred to a research center near Paris, and Vincent attended a boarding school in Paris. Vincent had skipped a couple of grades in the US. He was in the eighth grade then, and did very well at school.

Wu's outstanding achievements earned her many honors that year. She received an honorary Doctor of Science degree from Princeton University in June, together with Lee and Yang. She was the first woman recipient in the hundred-year history of Princeton. She was elected to the US Academy of Sciences, the seventh woman to be so honored.

Wu was promoted to full professor at Columbia University. Scientists believed that this was long overdue.

Wu did not forget the importance of conserved vector current in weak interactions while being busy with those activities. After a year of preparation, and several months of intense data collection and analyses, Wu, a student from Taiwan, Mo Wei, and a Korean student, Lee Rong-Gen, successfully confirmed the theory of conserved vector current in December 1962.

The experiment was done using the Van de Graaff accelerator at Columbia with proton, heavy hydrogen, and helium beams. The beta ray spectra were measured in the magnetometer spectroscopy fifty feet from the accelerator. The beta decay sources B-12 and N-12 were produced in the magnetometer.

Mo said that the three of them worked in tandem, collecting data non-stop. Wu started at eight in the morning, and worked until one or two past midnight.

Mo had graduated from Taiwan University with a major in Mechanical Engineering, and went to the Institute of Nuclear Science at the National Tsinghua University. In 1959, the founder of the institute, Sun Guan-Han, recommended him to study with Wu at Columbia.

Mo was tall, a straight talker, and had the Shandong (a province in northern China) no-nonsense personality. I flew from Washington to Roanoke, Virginia on March 24, 1990. He and his petite wife picked me up at the airport, drove to their home, went out for lunch, and sent me back to the airport. He told me many stories in three to four short hours.

Mo remembered that the neighborhood around Columbia University was not that safe, especially around one in the morning in winter. The

Wu in the laboratory of Columbia University.

Wu and other members of her research group in the low-temperature laboratory.

university was very cautious, and locked up all the buildings. The Pupin Laboratories had windows on the ground floor that opened to the street. A student found a key to the lock of a window gate, and Mo duplicated the key.

Around midnight when they were done for the day, Mo escorted Wu home in the nearby apartment, after they got out of the Pupin Laboratories through the window, instead of asking the campus guard for the key to the regular entrance. Mo would go first, jumping down onto the street. Wu then handed him a chair, climbed through the window, and stepped on the chair first, before leaping onto the street. Mo returned the chair and locked the window gate.

A campus guide saw them one night going through this routine. He watched in amusement, but stayed quiet. Mo smiled when recalling that Wu did not have enough exercise, and was wearing a *qipao*. She bruised her ankle so badly that she had to check into hospital after the experiment.

Mo worked in her laboratory from 1959 to 1963, and earned a Ph.D. He was an active high-energy physicist, teaching at the Virginia Institute of Technology and conducting research at DESY, Germany.

The annual meeting of the American Physical Society was held at the Hilton Hotel in New York City on January 26, 1963. Wu presented their results. As her experiments had always been well-thought-out and credible, the results were received positively and highly praised. This was her second prominent experiment at Columbia. The chairman of the physics department, Polykarp Kusch, who won the 1955 Nobel Prize in Physics for the accurate measurement of the magnetic moment of electrons, said that the conserved vector current experiment and the 1957 parity nonconservation experiment had the same elegant tradition of "clever conception, excellent execution, and great results".[13]

Wu was pleased with the achievement: "This result provided a more solid foundation to the universal Fermi interaction. It was so consistent with the theoretical explanation of a slight correction to otherwise 'pure' theory of conserved vector current. This experiment also provided strong support to the high-energy physics experiment in July 1962 proving the theory of two-component neutrino."

The high-energy physics experiment that Wu referred to was also done by physicists at Columbia University. T. D. Lee raised the problem in a coffee break, and Melvin Schwartz proposed an experiment to investigate if there was another type of neutrino. Lee and Yang wrote a theoretical paper discussing the importance of the neutrino studies. The paper listed eight "must-solve" problems, with the possible existence of a second type of neutrino as the first problem. This paper and Schwartz's paper on experimental design were published in the same issue of the *Physical Review*. The Italian physicist Bruno Pontecorvo, who was exiled to Russia after leaking atomic secrets, had also proposed a similar experiment earlier.

The experiments were done at the Brookhaven National Laboratory (BNL) and CERN, but the latter ran into some design problem. The physicists Schwartz, Lederman, and Steinberger, who performed the BNL experiment, won the 1988 Nobel Prize in Physics. Schwartz, who first proposed the experiment, but was resentful of the cutthroat competition in physics, and the large-scale, complex team structure, had already left physics and founded a computer company in Silicon Valley in California.

Indeed, pioneering works were frequently (unfairly) scrutinized in the highly competitive 20th century. Wu's experiment was a good example.

Ten years after the completion of Wu's experiment on conserved vector current, the physicist Frank Calaprice of Princeton University and his student B. Holstein claimed that they had found a "second-class current". In 1976 they published a paper, the result of which was quite similar to that of an earlier experiment by a Japanese physicist, K. Sugimoto. The well-known theorist S. Treiman of Princeton then wrote a long paper arguing that Wu's experiment was wrong and the second-class current was true. There was a little turf war going on between Princeton and Columbia.[15]

Wu learned of this controversy. She was very upset when she called Mo.[16] They were also unsure and uncomfortable as they did use an incorrect Fermi function when they conducted their first experiment. Some theorists started to consider ways to modify the Fermi theory if a second-class current indeed existed. Among them were I. Primakoff of the University of Pennsylvania and his student Huang Wei-Yan, now chairman of the physics department of Taiwan University. They wrote a paper concluding that theory would have to become very messy to accommodate the second-class current. The Japanese group repeated its experiment; this time the result was consistent with a theory without a second-class current.[17]

Wu asked several students to repeat all data analyses in the first experiment, this time using a computer. They found that, fortunately, the incorrect Fermi function that they used did not impact or invalidate their result. Wu wrote a paper in *Reviews of Modern Physics* in 1977 clearing up the confusion. She succeeded in upholding her reputation, but was still left with a bitter taste.

The conserved vector current experiment was another major achievement for Wu, after the parity nonconservation experiment. This work was the first experiment to demonstrate the intimate relationship between electromagnetic interactions and weak interactions. The relationship, and the unification of these interactions, became mainstream research in the following forty years, with fruitful results in both theories and experiments. Wu's experiment was done at a critical juncture.[18]

In addition, Wu and her collaborators started the double beta decay experiment in the mid-1960s, in a salt mine 2000 feet underground near Cleveland, Ohio.

Double beta decay originated from theoretical consideration. Physicists long agreed that a neutrino is emitted with an electron in beta decay. In 1928, the British physicist P. Dirac proposed a wave equation to describe the motion of particles with spin one-half. The Dirac wave function has four components organized into a vector. Two components represent spin (up and down); two represent energy (positive for matter– neutrino and negative for antimatter–antineutrino). A neutrino is different from an antineutrino.

Fermi wrote a paper entitled "Quantum Theory of Radioactivity" four years later with the same notion. In 1937, the young Italian physicist Ettore Majorana wrote a paper proposing a wave function of only two components (spin up and down). His mathematical deduction and conclusion was very original, and had great impact.[19]

Majorana was a genius. He wrote only nine papers in his life, the paper on the two-component wave function being the last one. In the following year he hinted to his friends that he was contemplating suicide, and disappeared on a sailing trip. Fermi called him "a rare genius". In the Italian physics community, Fermi was called the "Pope" and Majorana the "Apostle".[20]

In Majorana's theory, the neutrino and the antineutrino are the same thing called the Majorana neutrino; whereas in Dirac's theory the neutrino and the antineutrino are different, referred to as the Dirac neutrino.

If the antineutrino is considered to be identical to the neutrino, (second-order) double beta decay could be thought of as the process in which an antineutrino was (first) emitted and (then) reabsorbed, resulting in the emission of only two electrons. In 1957, the physicist M. Goldhaber and two collaborators at Brookhaven National Laboratory established that the neutrino has "helicity" −1, or is left-handed (spin is always opposed to velocity). Their result made the emission-then-absorption impossible. Majorana neutrino and neutrinoless double beta decay were invalidated.

Majorana's theory was based on parity conservation. Although neutrinoless double beta decay was invalidated, double beta decay experiments had implications for the dimension of the neutrino wave equation, the neutrino rest mass, and lepton conservation. These issues attracted Wu's attention.

Wu and her collaborators reported in 1970 on their search for the double beta decay of radioactive calcium (Ca-48) to titanium (Ti-48). They again reported in 1975 on the lepton conservation in the double beta decay of selenium (Se-82) to krypton (Kr-82). Both papers contributed to improving the accuracy of lepton number conservation and double beta decay.

Wu performed an experiment in 1970 investigating the "Einstein–Podolsky–Rosen argument", a study of fundamental philosophy in quantum mechanics. The three physicists questioned the completeness of quantum mechanics in 1935. Her 1970 experiment was a continuation of a 1950 effort, and she performed another one in 1975; they all provided solid experimental support for the completeness of quantum mechanics.

She did a lot of "exotic atom" work in the ten years between the mid-1960s to the 1970s. Exotic atoms were created by substituting for the electrons orbiting a nucleus with negatively charged (more massive) lepton or hadron, thus leading to atoms of higher energy and more compact in size, like helium. Exotic atoms have a short life, of the order of 10^{-10} s. The replacing leptons or hadrons either decay rapidly or get absorbed by the nuclei.

Both electromagnetic interactions and strong interactions could be investigated during the very short life of exotic atoms, leading to new knowledge in nuclear and particle physics.

The *Mössbauer effect* was discovered in 1958, and became a new tool in physics experiments. Wu developed techniques for Mössbauer studies at ultralow temperatures, and used the effect to study the structure of single crystals and other matter. In the 1970s, she also used the Mössbauer effect to study the iron ion (Fe^{2+}) in the protein of blood cells, and the hemoglobin structure and its relationship to sickle cell disease.

In addition, Wu had many achievements in instrumentation — the development of detectors. With her expertise on beta decay, she developed many radioactivity detectors, such as a window counter with special seal, a gas scintillation counter, and a semiconductor counter. She also built an ultralow-temperature nuclear physics laboratory at the Pupin Laboratories in the 1970s to study the Mössbauer effect, superconductivity, and test time reversal invariance.

Wu's contributions to instrumentation and techniques illustrated the reasons why she was a first-rate experimental physicist. A great experimentalist should understand current theories, and the strength and limitation of the apparatus, and be able to design an experiment making full use of the apparatus to test theories — deep insight into physics theory and rich experience.

Wu's experimental research gradually slowed down in the late 1970s. She took on many social responsibilities, and was concerned with other problems. She remained active until her formal retirement in 1980.

Many events occurred in the period after her parity nonconservation experiment in 1957. Her dear father died in Shanghai on January 3, 1959. As China and the US were in the Cold War, she was not able to go home for the funeral. They had not seen each other for twenty-three years, ever since she left home in 1936. Wu was understandably very sad, and wrote "My heart was breaking. Could not hold back my tears!" in a letter in 1959.[21]

Her elder brother, Chien-Ying had also died from illness in June of the previous year, and her mother, Fan Fu-Hua, died in October 1962. Wu was sad about losing all her family members. Her own family was rather peaceful, with both Luke and her busy with research. Their only child Vincent,

▲

For the 80th birthday of Isidor I. Rabi
(*fourth from right*), Wu (*fifth from right*)
celebrated with the President of
Columbia University and representatives
from other countries.

▲

Wu participated in the selection
process of the Westinghouse Science
Talent Search.

studied at the famous Bronx High School of Science, and then Columbia University, majoring in Physics.

Wu's achievements were properly recognized. She received numerous awards and honorary degrees, and served in many important positions.

Notes

1. This paragraph was based on "Wu Chien-Shiung — The First Lady of Physics Research", Gloria Lubkin, *Smithsonian*, 1971. Also an interview with Wu Chien-Shiung.
2. Telephone interview (between Taipei and Hong Kong) with C. N. Yang, June 21, 1994.
3. "History of Beta Decay", Wu Chien-Shiung, in *Trends in Atomic Physics — Essays Dedicated to Lise Meitner, Otto Hahn, Max von Laue on the Occasion of Their 80th Birthday*, eds. O. R. Frisch, F. A. Paneth, F. Laves and P. Rosband, Interscience, New York, 1959.
4. "The Discovery of Non-conservation of Parity in Beta Decay", Wu Chien-Shiung, in *Thirty Years since Parity Non-conservation*, ed. Robert Novick, Birkhauser, Boston, 1988.
5. Letter to Adina Wiens from Wu Chien-Shiung, December 18, 1957.
6. "Subtleties and Surprises — The Contribution of Beta Decay to an Understanding of the Weak Interaction", Wu Chien-Shiung, in *Five Decades of Weak Interactions*, New York Academy of Sciences, 1977.
7. Interview with Wu Chien-Shiung, February 12, 1990, New York residence.
8. "Subtleties and Surprises — The Contribution of Beta Decay to an Understanding of the Weak Interaction", Wu Chien-Shiung, in *Five Decades of Weak Interactions*, New York Academy of Sciences, 1977.
9. "Stories of Several Physicists", C. N. Yang, Lecture at Chinese Science and Technology University, June 6, 1986. Telephone interview with C. N. Yang, April 9, 1993.
10. Same as 9.
11. Same as 8.
12. *Newsweek*, p. 5, May 20, 1963. When Gell-Mann visited Taipei on April 12, 1995, he told the author that Wu had insisted that he and Feynman calculate the "correction terms" clearly before she would start the experiment.
13. *Columbia University News Release*, January 26, 1963.
14. *Columbia University News Release*, January 26, 1963.
15. Interview with Mo Wei, March 24, 1990, Blacksburg, Virginia residence.
16. Same as 15.
17. Telephone interview with Huang Wei-Yan, June 29, 1994, Taipei.
18. Interview with T. D. Lee, October 2, 1989, office at Columbia University. Interview with Mo Wei, March 24, 1990, Roanoke, Virginia residence.

19. Telephone interview (between Taipei and Hong Kong) with C. N. Yang, July 18, 1994.
20. *The Tenth Dimension*, Jeremy Bernstein, McGraw-Hill, New York, 1989.
21. Letter to Hu Shih from Wu Chien-Shiung, May 1, 1959.

Chapter 11

The First Female President of the American Physical Society

In October of 1964, at the MIT auditorium, an elegant, petite, middle-aged lady wearing a *qipao* rose slowly and then confidently addressed the audience with a slight Shanghainese accent: "I wonder whether the tiny atoms and nuclei, or the mathematical symbols, or the DNA molecules have any preference for either masculine or feminine treatment." Her statement received immediate applause from the audience.[1]

The applause at MIT was not the only applause she was to receive. Wu had been a top woman nuclear physicist in the world since 1940. As a female scientist, the struggles and difficulties she had faced for twenty years had made her especially sensitive to, and concerned about, the opportunities and privileges with regard to women in science.

Driven by this concern, Wu always encouraged her female students to devote themselves to their scientific research, and on many other occasions called for fair treatment of women in science, with her own personal experience as an example.

At that symposium entitled "Women and the Scientific Professions" at MIT, Wu had been invited as a principal speaker. She protected women's rights, and spoke against the deep-seated prejudice towards women in society.

In this talk, Wu started by quoting what Matthew Vassar (who founded Vassar College, a top women's college in Poughkeepsie, New York) said a hundred years ago: "Women inherit the same intelligence as men from our creator, thus they have the same right and responsibility to advance science and literature in the world."

Wu said that the founder of women's education had this progressive philosophy a hundred years ago, but sadly there had been little advancement since then. She asked: "What is the reason?"

She continued: "I sincerely doubt that any open-minded person really believes in the faulty notion that women have no intellectual capacity for science and technology. Nor do I believe that social and economic factors are the actual obstacles that prevent women's participation in the scientific and technical field."

She added: "The main stumbling block in the way of any progress is and always has been unimpeachable tradition."[2]

In a panel discussion later, a social scientist named Bruno Bettelheim told the story of a young Russian woman in engineering. He claimed: "She loves her job, embraces it like a mother. It is very different from men with a conquering instinct." His argument drew a firm but humorous rebuttal from Wu.

In the 1960s, the US had become a superpower with advanced technology and science, as well as a booming culture. The deep-seated prejudice against women in society had not correspondingly diminished.

Wu quoted a 1962 report that painted a vivid picture of women's status in science. The report was prepared by the American Association of University Professors, and discussed "Women in the Top Teaching and Research Institutes". The survey said that in the top ten private and public US universities, only 10% of the assistant and associate professor ranks and 5% in the full professor rank were women.

Wu also reported in "My Views on Women in Academia" in 1972 that in the top fifteen physics departments there were 760 male associate professors and above, but only 6 females.[3] There had not been much improvement by the end of the 1970s.[4]

She was surprised by the prejudice against women in the US when she first arrived at Berkeley. In 1936, she was supposed to just stop by Berkeley on her way to the University of Michigan, until she learned that the student center there would not even allow women to use the front entrance. She was very resentful and decided to stay at Berkeley.

Many years later, Wu still suffered from this prejudice in society, and was always sensitive to the cause. She said: "Women in America are really oppressed!"[5] In an interview with a newspaper in San Francisco in 1963, she pointed out that there were only 7% women in all sciences, and 3% in physics.

She said: " In America, there is a wrong image that all female scientists are boring old maids. It is men's fault ... In Chinese society, a woman is measured solely by her merit. Men encourage them to succeed, and they do not have to change their female characteristics in doing so."[6]

Wu pointed out: "US society and families unfortunately believe that science and some other fields are exclusively men's turf ... It is different in China. My father was an educator, and might have been ahead of his time. In the 1930s, Chinese society realized that we had to deploy all resources — collective talents of *both* men and women — if we want to catch up with the West." She added: "The West is ahead of China in science and technology, but not necessary in the effective utilization of human talents."[7]

Her own experience was very different. She was born into a family indifferent to gender. Her father was progressive, always encouraging toward his only daughter. The Ming De School, which he founded, was for girls. From a very young age, Wu believed that men and women were equal.

Whether at Soochow Girls' High School, the National China College, or the National Central University, she and her girlfriends never encountered discrimination as females. To the contrary, she received unusual encouragement and nurturing because of her outstanding talent and persistence. The principal of Soochow Girls' High School (Yang Hui-Yu), her advisor Gu Jing-Wei at Academia Sinica, and the positions and careers which her friends had achieved in China convinced her that women were treated more fairly there.

Wu, like other Chinese intellectuals, had expected an open, progressive society in the US. But the longer she stayed there, the more she realized that the US, perhaps heavily influenced by the Protestant philosophy, had deep-seated prejudice against women. It had negatively affected her career.

From 1944 to 1952, it had taken her eight years to be promoted to associate professor at Columbia University. Usually eight years was not that long for a career at an Ivy League school, but considering her unusual achievements, the promotion was long overdue.

Wu had encountered sex discrimination even earlier. After her graduation in 1940, and two years of postdoctoral research, she could not get a teaching position at UC Berkeley in spite of her outstanding achievements. The only reason was that there were no women Physics professors at the top research universities.

Her professor at Berkeley, Emilio Segrè, had said that he would never forgive the then chairman of the physics department Birge for not keeping Wu.[8] In his autobiography, Segrè also accused Birge of discriminating against women.

After a year each as an assistant professor at Smith College and an instructor at Princeton, Wu went to Columbia in 1944 as a senior scientist, not a teaching position.

Wu was well established by then. Her experiment (with Albert) on the beta ray spectra in 1949 greatly cleared up the confusion and the debate, and had a critical impact on this field. This, and the subsequent experiments, had made her an authority on beta decay.

But Wu was not promoted. The main reason was that I. I. Rabi, the most powerful member of the Columbia physics department, was an old-school physicist, with a stubborn opinion of women. He valued Wu, cared for her, but never raised the question of promoting her.

One story was that Lamb (Nobel laureate in 1955) proposed promoting Wu in 1951 and was turned down. He was so upset that he left Columbia in 1952.[9] Steinberger, who stayed in Columbia for 18 years (1950–1968), also had proposed Wu for a teaching position. He said that it was very difficult for a woman to get a teaching position in the 1950s. Rabi, being pragmatic, felt that keeping Wu as a researcher was good enough.

Employing a woman scientist had already made Columbia University a pioneer among its peers.[10]

Other Nobel laureates at Columbia, such as Townes, Steinberger, Lederman, and T. D. Lee, all agreed that Wu was not treated fairly. She was discriminated against as an Asian, but more so as a woman.[11]

H. Schopper, who was once the Director of DESY, Germany and later the Director of CERN, Geneva, had profound understanding of the situation of women in physics. He believed that there was an unfriendly culture for women in physics, although not outright restrictions.[12]

In a way, Wu had relatively good fortune compared to what had happened in the earlier days. Schopper, now the President of the European Physical Society, had worked with the Austrian physicist Lise Meitner in the 1930s. She told him that discrimination against women was open and public in those days: she could only work in the basement, use the backdoor reserved for the laborers, and had no office.[13]

Meitner had great achievements in physics. She discovered nuclear fission induced by slow neutrons in 1938. Her collaborator Otto Hahn won the Nobel Prize in Chemistry in 1945, but she was excluded. She did receive the prestigious Fermi Prize later, but people were still puzzled by the Nobel decision.

In 1957, Wu invited Meitner, then 80 years old, to visit Columbia University and to dine in the famous Rainbow Room. Concerned about the possible problem of incontinence in old age, Wu asked Meitner if she had to use the restroom. Meitner told her that when she was doing research in the basement of the Kaiser Wilhelm Research Institute, there was no lady's room as only carpenters were in the basement. She learned then to drink little water, and avoid using the restroom. She said: "I am well trained."[14]

As another woman physicist, and being in the same research area, Wu praised Meitner highly. She felt that Meitner was a real expert in radioactivity, perhaps even more so than Madame Curie.[15]

But, in terms of fame, Meitner was far behind. She was a female, and a Jew to boot, and never won the Nobel Prize with its associated glory.

Wu felt sorry for Meitner, but she was sort of in the same boat. She still had the feeling that her profound achievements in physics were not fully appreciated according to common gauges.

It was clear that three women physicists had made tremendous, and representative, contributions in the 20th century. They were Madame Curie, Meitner, and Wu.

In the early 20th century, Madame Curie discovered radium, and greatly advanced the understanding of radioactivity. She and her husband shared the Nobel Prize in Physics with Becquerel in 1903. Marie Curie also won the Nobel Prize in Chemistry in 1911, for the discovery of radium. With two Nobel Prizes, she was naturally world-famous.

Meitner continued with radioactivity research. Her most important work was the identification and understanding of nuclear fission. Although Hahn and his collaborator Strassmann had observed in Berlin in 1938 the strange phenomenon produced by uranium bombarded with neutrons, it was Meitner who correctly identified that it was nuclear fission — the

Wu in the Rainbow Room. She entertained the famous physicist Lise Meitner there in 1957.

splitting of a large nucleus into two. The identification was a result of her many years of research in radioactivity.

Before Meitner was exiled to Sweden in 1938, she had been working at the Kaiser Wilhelm Research Institute for more than twenty years. Her contemporaries, such as Einstein, Planck, Bohr, Born, Schrödinger, and M. Von Laue, all had high praise for her. Einstein called her "our Madame Curie" and believed that Austrian women were more talented than French women.[16]

Wu's research began with the discovery of nuclear fission. Her distinguished work on nuclear decay and fission in the 1940s and the 1950s earned her the nickname "Chinese Madame Curie" or "Madame Curie in the East". When she was a student in China and later in Berkeley, Madame Curie was her role model, and her classmates in China remembered her admiration for Curie.[17]

Her senior thesis advisor, Shi Shi-Yuan, at the National Central University had once worked for Madame Curie, so Wu and Curie were somewhat related professionally. Madame Curie died in 1934, and Wu went to the US in 1936. They never met each other. Wu later met many scientists in the US and Europe, and formed a somewhat different opinion of Madame Curie.

Segrè, her professor at Berkeley, met both Madame Curie and Meitner when touring Europe. He once said: "Gee Gee, the physics of Madame Curie was not that great. Your work is better!"[18] In his book discussing modern scientists and their discoveries, he made this assessment of Wu: "Wu's willpower and devotion to work are reminiscent of Marie Curie, but she is more worldly, elegant, and witty."[19]

In her own assessment, Wu felt that the public should not base the evaluation on fame, but substance. "Madame Curie was too tough, and had bad temper. Meitner was better in those respects."[20] She also said: "Chinese and Americans thought so highly of Madame Curie ... Chinese people think that calling me the 'Chinese Madame Curie' is an endorsement and honor, but I do not quite feel that way."[21]

Although Madame Curie, Meitner, and Wu had worked in a similar field, they belonged to a different space and time; there is no totally objective standard by which to compare their achievements. But the following comments are interesting.

Many top physicists agreed that Madame Curie, Meitner, and Wu had their peak achievements in the 1910s, 1930s, and 1950s respectively, and were the most representative, most distinguished women physicists in their eras in the 20th century.

The Chinese physicist C. N. Yang, who worked closely with Wu when she was doing the parity nonconservation experiment, and who had been in touch with her before and afterward, believed that the admiration and respect received by Madame Curie from her contemporaries was not as much as that accorded to Wu.[22]

Yang was well known for his broad interests and his insight into the historical development of physics. He believed that Madame Curie, Meitner, and Wu were the most distinguished women physicists of the 20th century. Madame Curie and Wu had perhaps made more contributions than Meitner.[23]

On the other hand, the other Chinese physicist T. D. Lee believed that Wu and Meitner were similar but Curie contributed the least among them. The reason was that in advancing the development of new theories, nuclear fission (Meitner) and parity nonconservation and the unification of weak and electromagnetic interactions (Wu) were superior.[24]

◀ Lise Meitner

▶

Marie Curie

Lee also believed that Madame Curie was indeed a great physicist, and very famous. But calling Wu the "Chinese Madame Curie" was not necessarily an endorsement. He said that it was an example of Chinese bad habits and the occasional demonstration of an inferiority complex.[25]

Leon Lederman, a colleague of Wu at Columbia, ranked Madame Curie and Wu as equal, with Meitner slightly below.[26] Norman Ramsey of Harvard believed that Curie created the field of radioactivity, and Wu confirmed parity nonconservation. These achievements put them ahead of Meitner, but all three deserved to win Nobel Prizes.[27]

Maurice Goldhaber of the Brookhaven National Laboratory felt that it is very, very difficult to compare scientists of different eras. He believed that Curie was most original, very tough, determined, and dedicated, but not as precise as Wu in doing experiments. Meitner was not as original as Curie, and not as precise as Wu, but her work was multifaceted.[28]

Robert Serber of Columbia believed that Curie was really a chemist, not a physicist. Meitner was a good physicist, but not at the same high level as Wu.[29]

Glenn Seaborg ranked the achievements of Madame Curie, Meitner, and Wu in that order. Seaborg and Serber both mentioned another woman physicist, Maria G. Mayer (1963 Nobel laureate), believing that her achievements were not as important as Wu's.[30]

These many physicists all knew Wu very well, and had broad insight and understanding of many areas of physics. Their assessments were not identical, but all agreed on her profound achievements in physics.

Valentine Telegdi of CERN jokingly said: "I believe that Wu was much better than Curie, as Marie had (was helped by) a talented husband."[31] Madame Curie always worked with her husband.

Wu rebutted this point in the "Women in Physics" panel discussion in 1971. She said: "We all knew that Pierre was talented. But it is also undeniable that Marie's talent and persistence achieved their first discovery. People do not quite understand that it was *her* experiment, and she invited Pierre to join her. Then her second Nobel Prize was awarded five years after his death."[32]

Indeed, Marie Curie attained a unique position as the very symbol of a successful woman scientist. But her achievements in science might not be as superior and outstanding as the public was led to believe.

In this sense, Wu deserved a fair and objective assessment of her outstanding achievements. Also, it might be too narrow and not respectful to rank her *only* against other women physicists, and not against all physicists.[33]

As a woman scientist, Wu faced the expectations of a scientist, a wife and a mother, and the associated conflicts of these roles in almost 50 years of her career.

Because of her personal experience, Wu had a firm belief in how to balance family life and scientific career. She used Marie Curie, her daughter Irene, Lise Meitner, and Maria Mayer as examples: "There were so few people in history who had made such tremendous contributions in such difficult times."[34]

She asked: "Why didn't we encourage more women to go into science?" She believed that we had few women scientists because girls were not interested in sciences, which in turn was a result of the different ways in which boys and girls were educated as teenagers. "Unfriendly social atmosphere and psychological hindrance are at the root of the problem."[35]

Wu pointed out that a similar situation existed in different cultures and in different places to various degrees. We had to improve the social atmosphere and eliminate the psychological hindrance. Parents, teachers, schools, social scientists, and the government had to work together.

In particular, tradition worked harshly against women scientists. Wu had obviously encountered this difficulty personally. There are many ways to reduce such conflict between family and research. Husbands and wives sharing the housework is one way, and providing professional daycare centers is another, in order to resolve the issue of the childbearing duty of young mothers.

Wu said that another major factor is "whether the husbands truly respect the interests of their wives, are considerate, and are willing to reduce the burden of their wives". She quoted the findings of social scientists that children would benefit the most with equal care from both parents, and men should be more devoted as fathers, enabling mothers to spend more time on their interests.[36]

Wu had a happy family life. Her husband, Luke Yuan, was very considerate, and supportive of her research. Luke was a high-energy physicist working mainly in the field of accelerators. They were in different fields, and Luke was not as distinguished as Wu. Their old friends said that Luke was a perfect husband, considerate and supportive, and took care of all details in all those years.[37]

Their friends were all impressed by Luke's modesty and consideration. Wu had a stronger personality, and was more confident and decisive, traits that had been built up from her earlier success. Luke would go along with her decisions, and was very complementary.[39] But she also maintained the traditional femininity, leaning on Luke at times: "Wait till Luke returns." In her old age, Wu would more often comfort Luke when he had a bad day or a short temper.

Luke would do more than his share of housework, as Wu would be in the laboratory from dawn till dusk. She would cook when she was free or had company. Luke was a good cook himself; he admired her cooking, listing "lion's head", chicken, sautéed vegetables, and *wonton* as representative dishes.[39]

Wu once said: "I always had a lot to read, letters to write and research matters to take care of. Not much time left to do housework. Fortunately, I have a very considerate husband, who is also a physicist.... I am sure he will be very happy if I let him return to work without constantly bothering him."[40]

She was too busy to spend much time taking care of their only son. Vincent had a nanny in Princeton and in New York. He attended a boarding school in Long Island in his first year in grade school. Luke worked at the Brookhaven National Laboratory then and stayed in their house in Long Island on weekdays. He sent Vincent to school every Monday morning and picked him up every Friday afternoon, then took the train back to New York City. They would wait for Wu in her laboratory to go back to their home near Columbia for the weekends, with Luke helping her and her students to finish their tasks.

Vincent returned to New York City after two years in Long Island and enrolled in a private school. He took the school bus in the early morning, and returned home after four in the afternoon. He learned to take care of himself and finish his homework. At times he would call Wu, begging her to come home, as he was so hungry.

Sometimes Wu had to work until midnight and Vincent would call again. She would comfort him, and brag about his ability to open a can of spaghetti for dinner the next morning.[41]

Vincent later worked at the Los Alamos National Laboratory in New Mexico. He recalled taking care of himself when growing up. He was used to simple food like chicken and canned spaghetti and "rather like it".[42]

Like other mothers, Wu was concerned about his schoolwork. She once told a woman student that the Mathematics teacher had taught Vincent that a finite number divided by zero would be zero. It was obviously wrong. She was very upset, and wanted to discuss the matter with the teacher.[43]

Vincent also remembered that his mother had to ask the school if he could take algebra a year in advance. But he said that his mother's concern was a "big picture" kind, she never checked up on him for every little thing, and always left him ample personal space for development.[44]

Wu had an acceptable arrangement, but she realized that a child would add a certain burden to anyone devoted wholeheartedly to research. She always advised her female students that they must find a good nanny to care for their children, and never give up their research.[46]

Wu was particularly close to her female students, sometimes acting like their mother. She would recruit women to her research laboratory. One of her students said that she had practically been "lured" to work for her.[46]

Wu's concern that a child would add a certain distraction was not only for women. A male Ph.D. student in the 1950s got married and fathered a child. Wu was very surprised to see the student and the child together walking in the street. She found it unthinkable to have a child while still a graduate student.[47]

Her dedication and achievement made her the "superwoman" type. But Wu maintained her femininity and her interests as a woman.

Wu's friend at the National Central University, the artist Zhang Qian-Ying, who lived near London, recalled that Wu would visit her whenever there was a meeting or something nearby. The Chinese ladies in London had a monthly social gathering. Wu would join the meeting if the timing was right. She was sociable, easygoing, and she never mentioned that she was a prominent scientist.

Wu was interested in discussions on cooking. She once argued that one must add cornstarch in the "lion's head" to make it silky and tasty.[48]

Wu was not that busy as a young student, and had time to do some shopping. As time went on, she had no spare time. Occasionally while riding out of town, she would ask Luke to swing by Fifth Avenue, so that she could do some "window-shopping".[49]

She did not pay much attention to makeup and dress. She always wore a *qipao*, a touch of lipstick, and occasionally went to the beauty salon. The architect I. M. Pei said that Wu never wore makeup and looked like an old-fashioned Chinese lady. She became totally modern once she was engaged in a conversation.[50]

As a good-looking woman scientist, Wu sometimes received unsolicited admiration from men, but she learned to handle it gracefully.

When she attended an international conference in Israel in 1957 right after her parity nonconservation experience, she was naturally the star of the meeting. An organizing staff was very attracted to the pretty Wu, and invited her to visit Israel for several months. Wu gracefully pointed to her collaborator S. Devons of Columbia University, another attendee, and said: "Ask him. I will come along if he agrees."[51]

She was also sensitive to her treatment by men. She once attended a committee meeting of the National Science Foundation in Washington DC with Robert Serber of Columbia University. A man addressed her as "Professor Yuan". She immediately corrected: "I am Professor Wu and Mrs. Yuan".[52] When providing materials to the Public Relations Department at Columbia University, she also indicated that she would prefer to be addressed as "Professor Wu", rather than "Mrs. Yuan".[53]

At the annual meeting of the American Physical Society (APS) in early 1973, it was announced that Wu had been elected as "Vice President Elect". According to the by-law of the APS, the Vice President Elect would automatically become the Vice President in a year, and the President in another year. Wu became the President of the APS in 1975.

With some understanding of the history of the APS, the election of Wu to be the President was a major event. The US scientific field had always been white-male-dominant. As many APS members were leaders in the science and technology development effort of national defense in the Second World War, the APS was a particularly influential organization. As an example, when President Reagan proposed a "Star Wars" defense system, the APS made public its objection and had the proposal greatly revised.

As she was a Chinese-American woman, the election of Wu as the President broke all those "long traditions" at the APS. She became the "first" woman president. The election received a lot of media attention. So far she had been the only Chinese-American serving in this position. Yang and Lee were nominated in recent years, but never elected.[54]

Wu received the support of the physics community as her achievements in physics were universally respected. She was at first not eager to serve, but was convinced after repeated urging by her colleagues.[55]

The APS was founded at Columbia University in 1899. The first three American Nobel laureates, Albert Michelson, Robert Millikan, and A. H. Compton, all served as presidents. Later presidents include Oppenheimer, Rabi, Fermi, Bethe, Wigner, Uhlenbeck, Weisskopf, and Seitz (who served as the Foreign Science & Technology Advisor in the Executive Branch in Taiwan). The early presidents were more ceremonial, with not much actual work; the presidency became an operational position in Wu's time.[56]

Wu became the President of the APS in 1975. She accomplished a great deal in a year. She was very diligent and was devoted to the job, just as when she was doing scientific research. Ramsey, who became the President of the APS three years later, said that her decisions in many scientific policies had impressed him very much. "She was a very good President."[57]

Panofsky, who was the President of the APS a year before Wu and the Director of the Stanford Linear Accelerator for many years, once said: "Wu was outstanding!"[58]

Wu recalled that she did not have much administrative experience at first. Panofsky advised her to delegate duties to many committees. She was successful in delegation, and greatly improved communications with the general public through frequent meetings with science reporters in the news media.[59]

Wu was very serious. She started an editorial in *Physics Today* addressing the problems faced by fundamental sciences, and solutions that the physics community should pursue. This effort mobilized the community, enhanced its effectiveness, and was well supported.

Wu, as the President of the APS, wrote a letter to President Ford on October 31, 1975. The letter explained the importance of research in fundamental science, and the need for long-term, steady funding support, and asked President Ford not to yield to short-term economic pressure and reduce the budget.

Ford personally replied. He reiterated his support for fundamental science, and re-established the Office of Science and Technology Policy (which was abolished by President Nixon in 1967) in the White House, advising the President on scientific matters.

In addition, Wu wrote a letter to V. Kirillin, the Chairman of the Russian National Committee on Science and Technology. The letter expressed her concern about the political interference in the scientific exchanges between Russia and the US, such as the case involving the Russian professor M. Azbel.[60] The Azbel matter was later resolved, and Wu was personally commended by an American physicist for her effort.[61]

Wu was pleased with this extracurricular activity. She said that Ramsey was scared to follow in her footsteps three years later. After an interview with her, the editor of a newspaper in the New York College of Sciences and Arts said: "She looked like a president in every inch of her body."[63]

Her brilliant achievements and lasting contributions resulted in a long record of honors and awards. The first one came in 1943 when she became the first woman instructor in the history of Princeton University. Fifteen years later, she was the first woman to be awarded an honorary Doctor of Science degree in the 100-year history of Princeton University.

She received the Research Corporation Award in January 1959 — again the first woman recipient. In addition, she received the Cyrus B. Comstock Award (given once every five years) from the National Academy of Sciences in 1964, as the first woman recipient among the past eleven winners.

Wu was also awarded an honorary LL.D. (the first woman) by the Chinese University of Hong Kong in 1969 and was the first Pupin Professor at Columbia University in 1972. She received the Scientist of the Year Award from the magazine *Industrial Research* in 1974, again as the first woman recipient.

Life magazine in the US published a special edition entitled "Remarkable American Women" in 1976 to celebrate the bicentennial of US independence. It profiled 166 distinguished women in US history. Wu, the author R. Carlson (*The Silent Spring*), the anthropologist M. Mead, and (the 1963

▶

Wu was awarded the Woman of the Year Prize (St. Vincent Cultural Foundation) by the President of Italy in 1980, and delivered a lecture after the banquet.

▲

Wu together with C. N. Yang (*front row, second from right*) and T. D. Lee (*first from right*), received honorary doctoral degrees from Princeton University in 1958.

Wu in the lecture hall where Galileo spoke, after receiving an honorary doctorate from the University of Padua in 1984.

Nobel Laureate) M. Mayer, among a total of sixteen women, were listed in the "Quest of the Mind" class.

An industrialist established the Wolf Prize in Israel in 1978. Wu was the first one in physics to receive this so-called "Israel Nobel". In 1981, she was awarded the "Woman of the Year" St. Vincent Medal, sponsored by UNESCO and awarded by the President of Italy. With the artist Georgia O'Keeffe, and six more people, she received the Lifetime Achievement Award in 1983 from Radcliff College, Harvard University.

The American Association of Physics Teachers held a "Women in Physics" conference in Atlanta, Georgia in 1990, and awarded Wu a "Special Commendation" for outstanding service in science and education.

Her achievements in science also earned her honorary titles. After the 1957 parity nonconservation experiment, she was called "the world's foremost female experimental physicist". Then her professor at Berkeley, Emilio Segrè, called her "the queen of nuclear physics". Wu was given the title "The First Lady of Physics Research" in the 1970s. Mrs. C. B. Luce, a one-time US ambassador to Italy, made the following interesting assessment: "When Dr. Wu knocked out that principle of parity, she established the principle of parity between men and women. People can no longer say that women are not capable of reaching the pinnacle in scientific research."[63]

E. Yost wrote a book entitled *Women in Modern Sciences* in 1959. Wu and Meitner (then eighty two years old) were the most representative ones among the eleven distinguished women scientists. Wu was recognized as the greatest woman physicist in the world at that time.[64]

Yost also pointed out that six of the eleven women scientists had a happy marriage. She remarked that she might have been the first to interview Wu, who was "very elegant, very warm, very straightforward". Yost (then 69 years old) said that she felt like "walking on a cloud" for several days after the interview. This charisma was highly unusual![65]

Based on her own achievements and experience, Wu tirelessly encouraged more women to go into science. She said that it would be a "terrible waste of intrinsic talents" otherwise.[66]

She said in a panel discussion on "Women in Physics" in 1971 that people might ask "Is there any benefit to the society if more women are encouraged to go into science?"

She added: "Men have always dominated the fields of science and technology. Look what an environmental mess we are in. They have pushed us to the brink of environmental disaster. Air, lakes, rivers, and oceans have all been polluted."

Wu quoted Rachel Carlson, who warned about the excessive use of the chemical DDT in her book *The Silent Spring*, A. Hamilton, who raised the problem of occupational health, and Dr. Frances Kelsey, who warned of the damage from thalidomide. She argued that the unique instinct of women, and their genuine concern, were exactly what society needed. She said: "The world would be a happier and safer place to live if we had more women in science."

Notes

1. "Reflections on Scientific Adventure", Rita Levi-Montalcini, in *Women Scientists: The Road to Liberation*, ed. Derek Richter, p. 111, Macmillan, New York, 1982.
2. Same as 1.
3. The report "My View on Women in Academia", Wu Chien-Shiung, 1972.
4. Same as 1, p. 113.
5. Interview with Wu Chien-Shiung, January 28, 1990, New York City residence.
6. *Newsweek* magazine, May 20, 1963.
7. Interview with *Herald American Boston* when Wu received an honorary degree from Harvard University, June 14, 1974.
8. Interview with Wu Chien-Shiung, October 26, 1990, New York City residence.
9. Interview with Jack Steinberger, August 9, 1989, CERN, Geneva.
10. Interview with Leon Lederman, March 27, 1990, office at the University of Chicago.
11. A similar feeling also from R. Serber and G. Feinberg of Columbia University.
12. Interview with Herwig Schopper, August 6, 1989, CERN, Geneva.
13. Same as 11.
14. Interview with Wu Chien-Shiung, April 24, 1990, New York City residence; *Fearful Symmetry*, A. Zee, p. 293, Macmillan, New York, 1986.
15. Interview with Wu Chien-Shiung, January 29, 1990, New York City residence.
16. *Nobel Prize Women in Science*, S. B. McGrayne, p. 48, Carol Publishing Group, New York, 1993.
17. Robert Wilson and other Chinese graduate students heard the same from Wu.
18. Interview with Wu Chien-Shiung, June 21, 1990, New York City residence.
19. *From X-Rays to Quarks*, E. Segrè, p. 260, UC Berkeley Press, 1980.
20. Interview with Wu Chien-Shiung, June 21, 1990, New York City residence.
21. Interview with Wu Chien-Shiung, June 21, 1990, New York City residence.
22. Interview with C. N. Yang, September 12, 1989, office at SUNY Stony Brook.

23. Same as 22.
24. Interview with T. D. Lee, October 2, office at Columbia University.
25. Same as 24.
26. Interview with Leon Lederman, March 27, 1990, office at the University of Chicago.
27. Interview with N. Ramsey, February 22, 1990, office at Harvard University.
28. Interview with M. Goldhaber, April 10, 1990, Brookhaven National Laboratory, New York.
29. Interview with R. Serber, March 14, 1990, New York City residence.
30. Interview with G. Seaborg, March 28, 1990, office at UC Berkeley.
31. Interview with V. Telegdi, May 22, 1990, CERN, Geneva.
32. *Physics Today*, p. 25, April 1971. The New York Physical Society sponsored the panel discussion on February 3.
33. Interview with M. Goldhaber, April 10, 1990, Brookhaven National Laboratory, New York.
34. The report "My Views on Women in Academia", Wu Chien-Shiung, 1972.
35. Same as 34.
36. Same as 34.
37. Interviews with Ren Zhi-Gong and Zhang Jie-Qian, September 26, 1989, Washington DC residence.
38. Similar views of friends, both Chinese and foreign, of Luke and Wu.
39. Interview with Luke Yuan, September 10, 1989, New York City residence.
40. *Christian Science Monitor*, September 10, 1965.
41. Similar opinions of Wu's students Noemie Koller and Mo Wei.
42. Telephone interview with Vincent Yuan (New York–New Mexico), September 10, 1990.
43. Interview with Noemie Koller, December 12, 1989, office at Rutgers University, New Jersey.
44. Telephone interview with Vincent Yuan (New York–New Mexico), September 10, 1990.
45. Interview with Noemie Koller, office at Rutgers University, New Jersey.
46. Similar opinions of her women students Noemie Koller and Evelyn Hu. Also interview with another woman student, Georgia Papaefthymion, March 13, 1990.
47. Interview with William Bennett, March 21, 1990, New Haven, Connecticut residence.
48. Interview with Zhang Qian-Ying, August 15, 1989, London residence.
49. Interview with Mo Wei, March 24, 1990, Roanoke, Virginia residence.
50. Interview with I. M. Pei, November 29, 1989, office at Pei Architects, New York City.
51. Interview with Samuel Devons, March 14, 1990, office at Columbia University.
52. Interview with R. Serber, March 14, 1990, New York City residence.
53. Letter to Mr. F. Knube, public relations office, Columbia University, from Wu, December 12, 1972.

54. C. N. Yang was nominated for Vice President Elect in 1990, and T. D. Lee was nominated in 1993.

55. R. Serber was the 1971 President of the American Physical Society. He recommended Wu and urged her to accept the nomination, March 14, 1990, New York City residence.

56. Interview with C. N. Yang, September 12, 1989, office at SUNY Stony Brook.

57. Interview with N. Ramsey, February 22, 1990, office at Harvard University.

58. Telephone interview with W. Panofsky, April 23, 1990.

59. Interview with Wu Chien-Shiung, July 31, 1990, New York City residence.

60. Letter to Vladimir A. Kirillin from Wu, September 12, 1975.

61. The letter was from Professor Edward A. Stern of the University of Washington, Seattle.

62. Interview with Wu Chien-Shiung, July 31, 1990, New York City residence.

63. *The Courant Magazine*, October 5, 1958.

64. *New York World Telegram and Sun*, April 14, 1959.

65. Same as 64.

66. *The Christian Science Monitor*, September 10, 1965.

Chapter 12

*L*ove of China

China has gone through periods of wars, turbulence, and prosperity in the 20th century. The historian Huang Ren-Yu observed: "China in the 1920s was like the Wei, Jin, Southern and Northern Dynasties (220–580 A.D.), having two competing governments in Guangzhou and in Beijing. It became prosperous in the 1990s, just like the Sui and Tang Dynasties (581–907 A.D.)." A 300-year history of turbulence was compressed into a 70-year transition in the 20th century! Modern China is no longer isolated from the West.[1]

This rapid change had no precedent. The massive population migration, and the conflict between the Communist Party and Kuomintang governments after 1949, had impact on the Chinese of the mainland and in Taiwan — the result was broken families, property damage, or physical abuse. Many overseas students might have avoided the direct impact, but also nonetheless suffered mentally from cultural adjustment and identity affiliation. Wu Chien-Shiung was one of the overseas Chinese. She had her own personal experience of this turbulent time.

She left China in 1936 by ocean liner for school in the US, fully planning to return home upon graduation, like others in that generation. The wars and political changes made her change her plan and stay in the US to do research and to teach. She married, had a family, and retired from Columbia University after 36 years of teaching. Her stay in the US was close to 60 years.

Wu had stayed reluctantly. As with other world-class achievers, her success built up her self-confidence, leading to her identification and pride in her Chinese heritage. In daily life, she maintained the Chinese customs and traditions, and often referred to the Chinese ways of doing things. During her long stay in the US, China was always on her mind in any situation.

The Sino-Japanese War broke out in the second year after her arrival in the US. Her dear uncle Wu Zhuo-Zhi visited her once in Berkeley; she otherwise had less and less contact with her immediate family. She worried about them and their safety, but could only focus on her study and research, hoping to return home and serve China upon graduation.

Wu had begun her research in the study of the radiation spectra in atomic decay. As soon as the atomic fission of uranium was discovered in Europe, almost all nuclear physicists rushed to investigate this new phenomenon and the development of the atomic bomb followed. During the research for her thesis, Wu obtained results that were later critically relevant to understanding the chain reaction in atomic fission. As a result, she was recruited in the spring of 1944 to join the Manhattan Project, which was developing the atomic bomb. As a foreigner participating in the secret defense project, Wu wanted to make a token effort to help the poor, war-torn Chinese.[2] The US had become an ally of China after the Pearl Harbor attack and the atomic bombs hastened the surrender of Japan, thus sparing the lives of many Chinese who would otherwise die in the battlefield. It turned out to be a real contribution to the Chinese people from Wu. Her original plan was to return to China as soon as the war ended. She once said that everybody in her teachers' generation had returned to serve the country. Her father would be so amused by the modern day excuse that Chinese students stay in the US because their wives are not willing to return.[3]

Wu could not return to China for many reasons beyond her control. When she graduated in 1940, it was near the peak of the Sino-Japanese War. Upon marrying Luke Yuan in 1942, the war had expanded to the Pacific Ocean. Eariler, Luke had been considering returning, without finishing school. Hu Shih, the Chinese Ambassador to the US, was against the idea, and convinced Luke and other Chinese students to stay put, finish school, and wait for future opportunities.

Wu and Luke moved to the East Coast. At the same time, she joined the Manhattan Project in 1944, and Luke joined RCA doing wartime research on radar. Japan surrendered in August 1945, marking the end of the Second World War. Wu and Luke again seriously considered returning to China. The National Central University offered both of them professorships, but suggested that they stay in the US for another year and purchase equipment necessary for setting up their laboratories. However, the conflict between the Communist Party and the Kuomintang government escalated, turning into a full-blown civil war.

For two years, they had correspondence with their families in China. Wu's father advised them not to rush home in view of the emerging turbulence. Wu became pregnant in 1946; their son was born in 1947. They became busy bringing up the child and doing research. The Communist Party realized total victory in 1949. Not wanting Vincent to grow up in a communist society, Wu and Luke postponed the plan of returning home.[4]

The Korean War started in 1950. There were anticommunist sentiments in the US in the 1950s, with the US and China becoming engaged in the Cold War. The US State Department imposed a virtual embargo on Chinese-American scientists, not issuing permits for them to visit China or return to the US. Wu and Luke gave up the idea of returning home. They both were naturalized in 1954, partly for the convenience of performing domestic research and in attending foreign technical meetings. In this period, Hu Shih, who had been the Chinese Ambassador to the US during the war, was invited to lecture at various US institutes. Having been Wu's teacher at the National China College, he had a profound influence on her both professionally and personally. At one time, Hu lived on East 81st Street in Manhattan, while Wu and Luke lived on West 116th Street. When Hu visited his dentist on the West Side, they had a chance to see each other.

When her dear friend Pauli visited Columbia University, Wu introduced him to Hu. Pauli had always been intrigued by Eastern and Chinese civilization. He had participated in many discussions with Luke about a book written by a relative of Luke. The book discussed the relationship between physics and the I-Ching. Pauli hit it off very well with Hu, a Chinese scholar whose knowledge was both broad and profound. They often talked for much longer than they originally intended.[5]

On February 4, 1957, the Eastern American Association of Chinese Scholars in New York City gave a banquet in honor of Wu, Lee, and Yang. The announcement early that year that Wu had experimentally confirmed the hypothesis of Lee and Yang (that parity conservation might not hold in weak interactions) had created much excitement all over the world. These three Chinese were in the spotlight, with extensive coverage in the media. This was perhaps the first time that Chinese had such exposure in international science.

Yang could not attend. With Wu and Lee attending, Hu was invited to introduce these three scientists. Hu said that Wu had been his student at the National China College, and he was so proud of her. He also said that both Lee and Yang had been educated at the Southwest Associated University, with Peking University as a member. The mathematician Yang Wu-Zhi, the father of C. N. Yang, was a good friend of Hu's at Peking University. Wu Da-Yu, a professor at Peking University, discovered T. D. Lee, and recommended him to directly transfer to the graduate school of the University of Chicago as a sophomore at the Southwest Associated University. Hu said that these connections painted a beautiful picture of the Chinese history of education.[6] After Hu, Wu was invited to speak. She said that she was indeed a student of Hu Shih, and appreciated his advice in the past; her experimental result was really nothing special, but a realization of a scientific method — "Boldly assume, but carefully prove" — that Hu had been advocating.

Wu recalled a lecture by Hu in which he told of the great philosopher Zhu Xi finding a fossil with the imprint of a frog in a mountain, and then deducing that the mountain must have been underwater thousands of years before. Hu said that China had science in the old days.[7] He had profoundly understood Chinese civilization. As a result of her early education from Hu, although Wu conducted research in Western science, she never lost her belief in Chinese civilization. Wu and Luke had the first opportunity for a "homecoming" in 1956, when they planned to conduct a lecture tour in Europe, Asia, and Taiwan. But, when Lee and Yang proposed the hypothesis of parity nonconservation in the weak interactions, Wu wanted to test the idea immediately, and canceled her trip.

Luke was working with the high-energy accelerator at the Brookhaven National Laboratory, and was later elected in the third round in 1959 as an academician at Academia Sinica. Mei Yi-Qi, President of Tsinghua University (Taiwan), invited him to advise on establishing the Institute of Nuclear Science, and on constructing a reactor in Tsinghua.

Luke went on the trip by himself. He attended a high-energy physics conference in Geneva, and he visited England, France, Italy, Egypt, and the

Tata Institute in India. He arrived in Taipei on July 19, and was welcomed by Mei Yi-Qi, Li Xi-Mou (Chairman of the Science Education Committee in the Education Department), and Li Kun-Hou of the Weaponry Department. The next day, the Taiwanese newspaper *Shin Sheng Daily News* reported on Luke's visit with the headline "World-Famous Genius in Atomic Energy". The article reported on an interview with Luke discussing various aspects of atomic energy. They apparently were quite ignorant of Wu, calling her Mrs. Yuan and spelling her name incorrectly.

Luke stayed in Taiwan for a fairly long time. He had extensive discussions with Mei about the construction of a nuclear reactor, and met President Chiang Kai-Shek at the Sun Moon Lake. The President was interested in Luke's opinion on whether Taiwan should develop its own atomic bomb. Strictly speaking, Luke's research was quite removed from the atomic bomb work. It was mainly on electromagnetic waves. He was involved in the development of radar during the war at RCA, and later at the Brookhaven National Laboratory in high-frequency electromagnetic waves. Neither effort was related to uranium fission and the atomic bomb.

But there was an atomic bomb expert in the Yuan family: Wu. She was involved in the Manhattan Project and had made critical contributions. Owing to her first-hand knowledge, Luke also had a rough knowledge of this area.

Based on this understanding, Luke recommended President Chiang Kai-Shek not to proceed with the development of an atomic bomb in Taiwan. He believed that it would consume an enormous amount of money, and he suggested channeling the resources to the peaceful application of atomic energy.[8] He recommended that Tsinghua University construct a nuclear reactor for education, and the production of radioactive isotopes.

Actually, many overseas Chinese scientists were against the development of an atomic bomb in Taiwan. The academician Ren Zhi-Gong, their six years senior, socialized with Wu and Luke frequently. He advised Luke not to touch the atomic bomb topic in Taiwan. When Ren discussed President Chiang's desire to develop a bomb with Wu Da-Yu, he gave the same advice.[9] Wu Da-Yu had consistently objected to the development of an atomic bomb in Taiwan, and this put him at odds with the military complex. This camp even circulated rumors questioning his loyalty.[10]

Following Luke's completion of his official business with Mei Yi-Qi, he toured the middle and southern parts of Taiwan for a month. In addition to Mei, the company included Li Xi-Mou, who later became the Director of the Atomic Energy Commission, and Sun Yun-Xuan, who was the Chief Engineer of the Taiwan Electricity Utility. One reason for Sun to come along was that there were no good hotels in Taiwan, and they often had to stay with the Taiwan Electricity Utility. Sun had since become a good friend of Luke and Wu.

The first time Wu went to Taiwan was in February 1962 to attend the Fifth Congress of Academia Sinica. It was her first homecoming trip after leaving China some 20 years before. It was also related to the fact that her teacher Hu Shih had been elected the President of Academia Sinica in 1957, and had returned from the US and lived there permanently. Before their trip, Wu and Luke had kept abreast of the political situation in Taiwan. When they left China, the Kuomintang government was facing multiple threats — an external one from Japan, and internal ones from the local war-lords and the Communist Party. In the 20 years since, it had won the Sino-Japanese War but lost miserably to the Communist Party. President Chiang Kai-Shek retreated to Taiwan and established a separate government. Overseas Chinese like Wu faced an identity crisis in this situation. During the Second World War, Wu loved the idea of a single China. In her mind, China should devote all its resources towards rebuilding the country after the war. Instead, China had a full-blown civil war between the Communist Party and the Kuomintang government. In a letter to an American friend, Wu was angry, and called such an act "stupid".[11]

Although Wu's parents and family did not move to Taiwan with the Kuomintang government, she found herself closer to Taiwan for several reasons. Her teacher, Hu Shih, had a close relationship with the Kuomintang government. He had written an essay explaining that the government had not declared war with Japan more readily as it was not sufficiently prepared and had to buy time. In the 1950s, because the US and China had developed an uncompromising attitude towards each other, the US government became suspicious of the communist government. Wu was not allowed to visit mainland China.

Wu was also involved in two incidents. The Taiwananese government had imprisoned Mr. and Mrs. Shen Zhi-Ming, the in-laws of the physicist

Kerson Huang Ke-Sun of MIT. On March 12, 1959, Wu, C. N. Yang, T. D. Lee, and Wu Da-Yu sent an urgent telegram to Hu Shih from the Institute for Advanced Study (IAS). The letter expressed their alarm and disappointment, and solicited his help for a fair investigation of this case.[12] Hu took the cause up immediately. He wrote a letter to Wang Yun-Wu (a prominent publisher and intellectual in China and Taiwan) on March 14 and had it delivered via telegram. Hu said that both the MIT and the IAS were world-class institutes, Yang and Lee were at the IAS, and Wu Da-Yu had visited the IAS the previous year. Hu emphasized that the Shen case could gravely damage the reputation of the Taiwanese government, and asked Wang to make sure that Chen Cheng (Vice President of Taiwan) understand the importance of this international incident. Hu requested that Mr. and Mrs. Shen Zhi-Ming be released on bond, and he was willing to act as a guarantor. Mr. and Mrs. Shen were soon released.[13]

In September 1960, Lei Zhen, the founder of the magazine *Free China Bimonthly*, was imprisoned. This was an influential publication, a liberal magazine, and a continuation of the "May 4" tradition. Hu once published articles there. Lei later organized an opposition party, and advocated the theory that "retaking mainland China is hopeless", and this had led to his arrest. The overseas Chinese were alarmed by the Lei case. Yang, knowing that Wu and Hu were dear friends, suggested that Wu get Hu's help. Hu happened to be in Seattle for a meeting of the Chinese–American Academic Cooperation Conference, and afterwards he visited New York City. Wu asked Hu to intervene. Hu promised to do so, saying; "I know him, and believe in his cause."[14] Hu returned to Taiwan on October 22. He raised the Lei issue with President Chiang, who obviously had a different opinion. Lei was not released, but was sentenced to a ten-year prison term. But Hu's public stand was complimented.

This led to the Taiwan trip on February 22, 1962. Wu was elected to Academia Sinica in the second round in 1958, and Luke the third round in 1959. They both attended the Fifth Congress of Academia Sinica. Taiwan was then rather isolated politically, as well as academically. Academia Sinica elected 81 academicians in 1948 on the mainland; only 9 fled to Taiwan, and 12 moved to the US in 1949. The Third Congress, in 1958, elected 14 new academicians, including Wu, Yang, and Lee. The Fourth Congress, in 1959, elected 9 new academicians, including Luke Yuan, Gu Yu-Xiu and

Liu Da-Zhong. There were four overseas academicians — Wu, Luke, the physicist Wu Da-Yu (who became the President of Academia Sinica in 1983), and the economist Liu Da-Zhong — attending the Fifth Congress. The credit for this rejuvenating event went to the President, Hu Shih.

It had only been a couple of years after Wu had completed her experiment confirming parity nonconservation, and she was working on several important experiments. She was influential in the international physics community, and at the peak of her career, creating a lot of excitement in Taiwan. Before the Congress, there were three meetings of Academia Sinica in Taiwan, and President Chiang met with them, as always. The Fifth Congress, with four overseas academicians attending, including a world-famous woman scientist, caught his special attention. When Hu had lunch with President Chiang on February 8, to report on the forthcoming Fifth Congress, the First Lady, Soong May-Ling, told Hu. "Be sure to bring her to see me."[15]

Wu and Luke arrived at 2:50 p.m. on February 22 in Taipei, and received a very warm welcome at the airport. The media focused on Wu. The important newspapers, like *Shin Sheng Daily News*, *Central Daily News*, *Credit Newspaper* (now *Reading Times*), and *United Daily News*, ran special reports the next day. The reports were accompanied with photos of an elegant Wu wearing a dark colored *qipao* and a pearl necklace. Wu and Luke were escorted to the Queen's Hotel, checked in, and rushed to visit Hu in Fu-Zhou Street. Wu Da-Yu, who had arrived earlier that same day, was already there. They had known each other before in the US, and were happy to meet again. Wu jokingly told Wu Da-Yu: "Since you were a student of Gu Yu-Tai, and he and I in turn were students of Hu Shih, you are a generation younger than me, and should address me as 'Aunt'."[16]

Wu gave two lectures: one at Taiwan University, discussing her parity nonconservation experiment, and another at Tsinghua University, on the Mossbauer effect. Luke was invited to lecture on "Recent Developments in High-Energy Physics" at Taiwan University. In the morning of February 24, the Fifth Congress of Academia Sinica elected seven new academicians after three rounds of voting, including Ren Zhi-Gong. There was a banquet at five in the afternoon, with Hu Shih and Li Ji speaking. Afterward, Hu asked Wu to speak, but she deferred to Wu Da-Yu, who spoke about the development of science in Taiwan. Then as Hu continued, he stopped briefly to give the audience time to partake in some refreshments. It was 6:30, and as Hu was mingling with the attendees, he suddenly turned pale, falling to the floor. Hu Shih, the most influential scholar in modern China, had died of a sudden heart attack.[17]

Before her return, a newspaper in Taiwan had reported that Wu's main reason for the trip was to visit Hu, and Hu had told the media how much he admired his student. His sudden death in her presence saddened her tremendously.[18] During the viewing the next day in the funeral home, she was trembling beyond control. Luke also had tears in his eyes on the 27th of February.[19]

Under such sad circumstances, Wu and Luke still had to deliver their lectures — Wu at Tsinghua University on the 27th, and Luke at Taiwan University on the 26th. They had originally planned to return to the US on the 28th, but were convinced by their friend Sun Yun-Xuan to rest for two days at Sun Moon Lake. They flew to Taipei and then on to the US on March 1, ending a totally unexpected and sad journey.

The next trip back to Taiwan was in July 1965. Wu received the first Achievement Award established by the Chi-Tsin (a cement company in Taiwan) Cultural Foundation. This trip received media attention in Taiwan, as well as a special report in *The New York Times*. After seeing the latter report, the President of Columbia University wrote a letter of congratulations stating. "This award represents the recognition of your scientific achievements from your fellow countrymen. You must be exceptionally pleased."[20]

Wu and Luke arrived in Taipei on July 15, after a week in Hong Kong. The newspapers published long articles about Wu, with photographs from the time when she was a young girl, and articles written by her old friends from the National Central University days — the artist Sun Duo-Ci and the educator Wu Zi-Wo — as well as many reports on her scientific works and family life.

Wu and Luke arrived in Taipei around noon, checked into the Yuan Shan Hotel, had lunch, and drove to the Hu Shih Memorial in Nangang to pay their respects. For Wu, visiting the memorial might have been more important than receiving any award.

Wu placed flowers at the grave and meditated there, immersing herself in sad memories of her beloved teacher. She then visited the Hu Shih

Memorial which held many of Hu's books and other items. Hu's 1936 letter to her was there. He wrote it while waiting for the ship in San Francisco for his return trip to China, and she was a graduate student at UC Berkeley. The letter, full of encouragement, was very valuable to her. Wu had found this 29-year-old letter a short time before her trip. She mailed it to Mrs. Hu, who immediately arranged to have it displayed before her visit. The memorial also displayed a certificate of her 1958 Research Corporation Award, which she had dedicated to Hu Shih. She stayed in Nangang for two hours.

Wu received the Chi-Tsin Award, and gave several lectures. In addition, she enjoyed very much the many private reunions with her old friends in Taiwan, more so than the official and political meetings.[21] An elaborate ceremony was held to present the Achievement Award in the Kuo Bin Hotel on July 21. Wu received a certificate and a gold medal with a picture of Confucius lecturing on one side and a picture of Lei Zu (the first woman scientist in ancient Chinese history) on the other. There was also an award of US$10,000.

After the ceremony, Wu presented a lecture on "Strengthening Scientific Education; Developing Scientific Research". She discussed the content and the process of her parity nonconservation experiment, and the conserved vector current experiment. The latter had been completed just one year earlier to test the hypothesis put forward by Feynman and Gell-Mann. At the end of her lecture, she emphasized how scientific research and scientific education should complement each other. Then she said: "I have a simple life with few living expenses. I would like to donate the award money to the cause of encouraging education in Taiwan. Isn't it consistent with the goal of the Chi-Tsin Culture Foundation to 'take it from the society and spend it in the society'?" The Board Chairman of the Chi-Tsin Culture Foundation, Wang Yun-Wu, applauded the donation, and said: "This way of taking from culture, and reinvesting it, is great. It will have a profound and meaningful influence on Chinese culture."

As Wu had many good friends living there, she cared much about Taiwan. But she had reservations about the Kuomintang government, as indicated by the two political incidents mentioned before. This was also consistent with her personality and belief. She once told a reporter: "My main

Wu and her teacher Hu Shih (*left*), with the physicist at Columbia T. D. Lee (*right*), in New York in 1957.

Wu in front of the burial memorial of Hu Shih in Nangang.

purpose of the trip is to visit the Chinese youth there. They are hopeful and have a bright future."[22] According to an overseas scholar, she was rather reluctant to meet President Chiang.

Wu and Luke nevertheless attended a reception for overseas scholars given by President Chiang Kai-Shek and First Lady Soong May-Ling at the National Defense Research Institute in Yang Ming Shan on July 17. In a private meeting, Wu formed an opinion of Soong that differed from the exciting one she had when she first listened to Soong's speech in the US capital.

Wu said that Soong looked very much like a person who had grown up in Shanghai, and they conversed in the Shanghai dialect. Wu apparently did not quite like Soong, and was unwilling to give details of their conversation, just saying that Soong was rather class-conscious.[23]

Wu believed that President Chiang Kai-Shek (Chiang Senior) was a nationalist, but that he deferred to Soong in all things international, thinking so highly of her.[24] Chiang Senior arranged to meet with Wu and Luke on both their trips to Taiwan, and raised the question of developing an atomic bomb. Wu said that such development would be too expensive, and not meaningful without supporting installations.[25]

Wu stayed in Taiwan for 14 days, receiving the award, lecturing, and touring along the new cross-island highway in the south, and returned to the US on July 28. At the airport, Luke read a written statement that Wu had written. In it, Wu thanked the public for their hospitality, and expressed their admiration for the many officials, academics, and students struggling in a tough environment. They pledged to help improve their lives. Chiang Ching-Kuo, the Defense Minister, visited Wu and Luke at their hotels on both trips the couple made to Taiwan. On their second trip, when Chiang saw them off at the airport, he related how pleased he was that six of the seven winners of the government-funded, foreign-study scholarship examination were Taiwanese, it was a sign of great improvement in local talent. His insight and inclusiveness impressed Wu and Luke.[26]

Wu and Luke had stayed in Hong Kong for a week before going to Taiwan. It was there that Wu was reunited with uncle Wu Zhuo-Zhi and younger brother Chien-Hao, her only surviving family members. Because of her uncle's love for her, he had funded her study in her first year at

Wu received the Achievement in Science Award from the Chi-Tsin Culture Foundation in Taiwan in 1965. She met President Chiang Kai-Shek, his wife Soong May-Ling, and his son, Chiang Ching-Kuo.

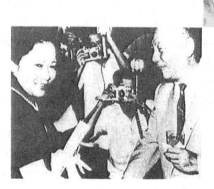

Berkeley, and had visited her there almost 30 years before. The meeting in Hong Kong was both emotional and moving.

Wu's elder brother Chien-Ying had died in 1958, followed by her father in 1959, and her mother in 1962. Her uncle had fallen in an accident several years before, when trying to avoid a dog darting out from nowhere while he was riding a bicycle. He was crippled. US and China were in the Cold War then, and the US State Department forbade any citizen to travel to China. Wu was helpless. Actually, through the high-energy physicist Zhang Wen-Yu, China had sought to have Wu and Luke return to their homeland permanently. The embargo[27] effectively forced them to stay.

In 1971, while competing at the world championship in Japan, the Chinese table tennis (ping-pong) team invited the American team to visit China. This was later called ping-pong diplomacy. 21 years of disengagement and isolation between China and the US was softening. The US National Security Advisor, Henry Kissinger, secretly visited China, and arranged the state visit of President Nixon the following year.

Among the famous overseas Chinese scholars, C. N. Yang was the first to visit mainland China. He noticed the easing of US–China relations in the spring, but also realized that the opportunity to return home might be lost again for some unforeseen reasons.[28] He toured China for four weeks, in July 1971. Afterward, at many conferences, Yang reported on his trip, and praised greatly the success and progress of China.

The next year, Yang traveled again to visit relatives. T. D. Lee, and also a group of scholars headed by the physicist Ren Zhi-Gong, visited China. These visits generated much discussion, and Ren especially urged Wu and Luke to take a look there.[29] This led to their trip in 1973.

Both Wu and Luke had left China in 1936, a good 37 years earlier. They arrived in Guangzhou on September 22, 1973, and from there they traveled to Shanghai, and then to her hometown, Liuhe. The old saying "Leave home as a youngster, but return as an oldster" accurately describes their feelings. In Wu's case, both of her parents and her elder brother had died. The uncle and the younger brother they had reunited with in Hong Kong had been tortured to death during the Cultural Revolution. The tombs of Wu's parents had been destroyed. Her painful experience was also a reflection of the turmoil and sacrifice of the Chinese people in the years preceding her return.

The 1973 trip lasted 53 days. Wu and Luke went to Hangzhou, Tianjin, Beijing, Luke's hometown (Anyang) in Henan, Zhengzhou, Luoyang, Xi'an, and Yan'an. They toured Kunming, Changsha, and Guilin, returned to Guangzhou on November 12, and left on the 13th.

They met many old friends in this long trip, and experienced the peculiar atmosphere that existed near the end of the Cultural Revolution. Many had suffered badly, but could not show their feelings in public. Luke met his sister Chia-Zhi in Tianjin, but they had to hold back their tears until they were safely inside Luke's bedroom in the hotel.[30]

Wu and Luke met with Chinese Premier Chou En-Lai in the Great Hall of the People in Beijing. Chou was a graduate of Nan-Kai High School in Tianjin. His progressive thinking and his visionary sight were greatly admired in China, and also respected by the overseas Chinese intellectuals. The meeting started with dinner at 6 o'clock on October 15, and lasted until midnight. It was their first and only meeting (Chou died in 1976). Wu and Luke were deeply impressed.[31] The meeting with Chou was held inside Anhui Hall in the Great Hall of the People. Chou told them that Anhui Hall had been chosen because Anhui Province lay between Jiangsu Province (Wu's hometown) and Henan Province (Luke's hometown). Such consideration earned their appreciation.[32]

Chou also sincerely apologized that the graves of Wu's parents had been destroyed, something beyond his control. In addition, he showed them treatises with Russia reclaiming Chinese land that the Kuomintang government had signed away.[33] The immense knowledge and the classy manner of Chou En-Lai as a great statesman impressed Wu and Luke to no end.[34]

In the following years, Wu and Luke returned to China several times to lecture. Near the end of the 1970s, China considered building a high-energy proton accelerator in Beijing. Wu, Luke, and the prominent high-energy physicist Zhang Wen-Yu expressed their support for this project but it was later canceled and replaced with an electron–positron accelerator. During their China trip in 1977, Wu and Luke were joined by their son, Vincent, and daughter-in-law, Lucy Lyon (they were married in 1974), who had flown in from Switzerland. This time they traveled to the Northeast Territory, Xi'an, Yunnan, Hangzhou, Shanghai, Suzhou, Kunming, and Shangri-La. They covered all four corners of China. It was the first trip for

Wu and Luke met with
Chou En-Lai in early
1973 in China.

Wu and Luke met with the President of Peking University, the famous physicist Zhou Pei-Yuan (*right*).

Wu and Luke met with the famous Chinese physicist Zhang Wen-Yu (*left*).

Vincent to return to the home country of his parents, and it motivated him to learn the Chinese language.[35]

Vincent was born in the US. He spoke French fluently, but did not speak Chinese. Wu was worried: "It is strange. He can master French, but has forgotten Chinese. My friends have the same problem. All their children have forgotten the Chinese language."[36]

In an interview the previous year, when Vincent was a 15-year old at the Bronx High School of Sciences, Wu mentioned: "I don't know if Vincent will become a scientist, but I know for sure that he is unfortunate in knowing so little of the Chinese language."[37]

But Wu never forgot her Chinese roots. Her external appearance — dress and manner — was very Chinese, and her internal spirit was even more so. She had profound understanding and confidence in Chinese culture. Her understanding, appreciation, and confidence regarding Chinese culture began with her upbringing, especially from the early teachings of her father. She said that her father had never studied abroad but was more progressive than most, and that she had met only a few people comparable to her father in those many years.[38] When she was a youngster, her father suggested that she study Chinese classics such as *A Selection of Classical Chinese Essays* and *The Analects of Confucius*. She always felt that there were profound principles in these old texts.[39] Wu had later gained real understanding of Chinese culture at school, especially after taking the course "History of Chinese Civilization" taught by Hu Shih at the National China College. This solid learning helped her maintain her roots as a Chinese, even though she spent most of her life in the US doing Western physics.

Wu recognized the cultural differences early on. When she learned of the practice of discrimination against women at the University of Michigan (student center), and the policy at the Ivy League universities (no women professors), she realized how progressive and advanced Chinese culture and society were. On many occasions, she publicly criticized sex discrimination in the US, saying that there had been equality in Chinese education for many years.[40]

Conversely, Wu realized that there were lingering bad habits in the overseas Chinese community. She went to the San Francisco Chinatown for the first time, attending the wedding of the brother of her friend Eda, an overseas Chinese. The wedding banquet did not start until 11 o'clock at night, after all the guests had slowly arrived![41]

While busy with her research and teaching, Wu was very involved with the Chinese community in the US, lecturing at or receiving awards from many Chinese organizations. When the first group of Chinese scientists toured the US in November 1972, they visited SUNY at Stony Brook (the tour was sponsored by C. N. Yang) and her nuclear laboratory at Columbia University.[42] Zhang Wen-Yu, who knew Wu very well at Princeton, led the group. He had worked with the physicist John Wheeler at Princeton. He was credited with discovering "Zhang radiation".

When Wu was the President of the American Physical Society (APS) in 1975, another group of Chinese scientists toured the US. Lu Jia-Xi, the Director of the Chinese Academy of Sciences and an old friend of Luke at Caltech, led the group, which included the solid state physicist Huang Kun. The visit was a big success with Wu's help from her position and influence in the APS.[43] President Chiang Kai-Shek died while the Chinese group was visiting. Ren Zhi-Gong saw the news in the hotel, thought that it was important, and rushed to tell Wu. She was quite indifferent.[44]

In 1984, China and the US resumed full diplomatic relations, and Zhao Zi-Yang (the Premier of China) officially visited the US. In New York City, Wu spoke at the banquet welcoming Zhao, and praised highly the new Chinese "open-door" policy and economic developments aiming at the improvement of living standards.[44] The government and some people in Taiwan were not pleased with these prominent overseas Chinese visiting the "illegitimate" mainland China. Some in the Legislature College even suggested canceling their title of academician. Fortunately, President Chiang Ching-Kuo objected. Wu and Luke stopped going back to Taiwan after 1965, to avoid this unpleasant situation.

Academia Sinica held a congress every two years. The President would visit foreign countries and hold meetings with overseas academicians in the off year. Wu and Luke did not return to Taiwan, but attended all these overseas meetings. At the meeting of overseas academicians in 1981, they proposed to President Qian Si-Liang a program for studying synchrotron radiation in Taiwan. It was the beginning of a closer relationship for the next ten years. Wu and Luke visited Taiwan in 1983. Pu Da-Bang, a physicist at UC Riverside, helped organize the trip. Pu's

At the Overseas Academicians' Conference of Academia Sinica in 1963. Front row, *from left to right*: Wu Da-Yu, Chiang Ting-Fu, Wu Chien-Shiung, Wang Shi-Jie, Wang Jing-Xi, Li Shu-Hua, He Lian. *Back row, from left to right*: Wang Shi-Jun, Bo Shi-Yi, Zhu Lan-Cheng, Cheng Yu-Huai, Yang Lian-Sheng, Luke C. L. Yuan, Li Jing-Jun, Zhou Wei-Liang, Ku Yu-Xiu, T. D. Lee, Chiang Shuo-Jie, Liu Da-Zhong.

father, Pu Xue-Feng, was a well-known political scientist, and served as an government official. After he finished high school (one affiliated with Taiwan Normal University), and graduate study in the US, Pu Da-Bang was rather active in the community of high-energy physics experimental-ists. He was energetic, and duty-bound. In the late 1970s, his career now stable, he devoted himself to the advancement of science in Taiwan.

There was the Atomic and Molecular Science Symposium, the first international gathering in Taiwan, in August 1979. Pu Da-Bang was the "midwife" of the conference, and one of the speakers was Li Yuan-Zhe (who later won the Nobel Prize in Chemistry.) In 1993, Li wrote an article reminiscing about his late friend Zhang Zhao-Ding, who was the Planning Director of the Atomic and Molecular Science Research Center in Academia Sinica. He had a vivid description of Pu Da-Bang:

"In the spring of 1978, I visited China with a group of chemists from the US Academy of Sciences. A couple of months after my return, I met Pu Da-Bang, who was visiting from UC Riverside, in the university dining room. We had heard of each other, but had never met. He asked about my observations in China, and we discussed many issues on the development of science and technology in Taiwan. The discussion was really exciting. We discovered what we had in common: a desire and dedication to advance science in Taiwan. Pu Da-Bang was truly competent and well-connected. He organized the Atomic and Molecular Science Symposium in Taipei in 1979. This successful conference attracted many distinguished scientists, both local and foreign, and helped prepare for the eventual establishment of the Institute of Atomic and Molecular Sciences in Academia Sinica, as well as the Taiwan Synchrotron Radiation Research Center."

Wu had not been to Taiwan for 18 years. Although she had retired three years before, she was still very well respected in the international nuclear physics community. Unfortunately the underlying hostility toward scientists who had ever set foot in mainland China made her visit to Taiwan awkward. Pu Da-Bang exercised his political skills. When he learned that Wu did not object to visiting Taiwan, he contacted Qian Si-Liang, the President of Academia Sinica, and suggested a formal invitation from Academia Sinica. Qian agreed. Pu and Wu worked out the details, resulting in her trip in March 1983.

Wu attended the Second Atomic and Molecular Science Symposium and talked to students at the National Central University, her alma mater. It was rare then to have a visit from scientists of international status. Her presence created a great deal of excitement in Taiwan. Nonetheless, the stories about Taiwan that Wu had heard were not very flattering. Returning from a visit to the National Central University, she asked skeptically how much a woman laborer earned in a month.[46]

Wu was instrumental to the progress of the synchrotron radiation program in Taiwan. People were debating whether they should build a synchrotron. Pu arranged for Wu and Luke to meet with President Chiang

Ching-Kuo. Wu and Luke did so, and convinced the President to support the project. The synchrotron was completed in ten years.

Wu and Luke went back and forth to Taiwan for the next ten years. Wu was a member of the Planning Board of the Synchrotron Radiation Program, as well as on the Technical Advisory Board. Luke was an executive member of the Planning Board. Wu's health had suffered from the hysterectomy she had undergone in 1985, but she still undertook these long-distance, grueling trips, making five round trips a year at the peak.

In 1988, Wu made a special trip to Liuhe to attend a celebration of what would have been the 100th birthday of her father. She donated part of her savings and prize money from awards to establish the Wu Zong-Yi Memorial Foundation, with an annual interest income of 60,000 yuan (Chinese dollars, approximately 8,000 US dollars at the exchange rate of eight yuan to one US dollar). Part of the interest income would be used to provide scholarships and build the library, and part would be used to improve teaching or for continuous education of the teachers at the Ming De School.[47]

Wu Zong-Yi had founded the Ming De School as a primary school. After several expansions, it grew to have classes from kindergarten to high school. The Southeast University Architecture School, a spin-off from the National Central University and part of the team constructing the Great Hall of the People, volunteered to design and build the expansions. Students started to learn English in the third grade, and computer use in the fourth. Wu bought many personal computers from Acer of Taiwan. These improvements helped children in this small village to finish high school.[48]

Wu adhered to the traditional teacher-student relationship in China, maintaining her respect for Hu Shih, who in turn took good care of her as a student. She did not care for some young physicists who thought too highly of themselves and harshly criticized their own teachers to no end.[49] Wu also took care of and nurtured young students at Columbia University — quite different from the typical "self-centered" great physicists there.[50]

She had a special affection for Chinese students. One of her students from Taiwan brought a girlfriend along to her party. She "approved" of her and encouraged them to get married.

The student graduated from Columbia, but was apparently not that successful in research. He joined a religious cult, and committed suicide by cyanide poisoning. Wu was very sad. After the funeral, she went to the Chinese Provincial Association, which he belonged to, and stayed for an hour trying to find out the cause.[51]

She was particularly fond of two students from Taiwan. One was Mo Wei, who did the conserved vector current experiment with her. Mo became an active high-energy physicist teaching at the Virginia Institute of Technology. Li Wo-Yan was the other one. He was energetic and active in politics. Li was once arrested for joining a study group at Taiwan University. He worked for Wu at Columbia but also was one of the leaders of the Diao Yu Tai Movement in 1970. The People's Republic of China (mainland China) replaced the Republic of China (Taiwan) in the United Nations that year, and Li was among the first group of five students invited to visit China.[52]

Li's trip was highly secretive. There was a rumor that when he went to China, he hung his jacket on his chair on purpose and arranged to have his car moved every day. He was Wu's postdoc but she had no idea that he was gone.[53] The trip became public only after his return. His passport from Taiwan was suspended. He gave up research and worked for the United Nations. Wu thought highly of Li, saying that he was smart and competent, and she maintained contact with him. When Li retired from the UN Environment Protection Agency in Nairobi in 1994, he invited Wu to Africa. Wu could not make the trip because of poor health. Li had remained on a blacklist in Taiwan for many years until 1993.

Wu mourned the disappearance of Chinese tradition. She admired I. I. Rabi of Columbia, who was maintaining the Jewish tradition very well.[54] Rabi was the most prestigious physicist of Columbia's physics department. After retirement, he often walked around chatting with various people, and they addressed him simply as "Uncle Rabi", instead of "Professor Rabi". Wu liked this democracy and informality. By comparison, she was uneasy about sitting with high officials in China, believing that the distinction was not a "socialist" way.[55] Professors in Taiwan were very class conscious too.[56]

Wu admired the talents of the US founders, and the prosperity of the country, saying that the US was quite generous to the immigrants.[57] She discussed science in China, referring to the astronomer and mathematician Zu Chong-Zhi of the Han Dynasty, and the medical scientist and the author of *Compendium of Material Medica*, Li Shi-Zhen of the Ming Dynasty. She mentioned the marble statues of these two scientists standing at the entrance to the University of Moscow.[58] Wu thought that it was really embarrassing to have the *Briton* Joseph Needham write the history of science and technology in *China*.[59]

Wu never believed that "only foreign countries have a full moon", and once said that many China experts in the US were actually quite ignorant.[60] Like many other Chinese experiencing the seismic changes in recent history, she had a big struggle with her identity and affiliation with different cultures. A friend of hers, the architect I. M. Pei, said: "Chinese remain Chinese, but gain a world-view." Both Wu and Pei had lived in the US for over half a century. Wu had a world-view, but remained Chinese at heart. Her sitting room had paintings by several famous Chinese painters, and many old Chinese texts on the bookshelf.

When Wu conversed with her dear American friends, she might revert to speaking Chinese.[61] She would do so unconsciously when she became tired working in the laboratory.[62] Wu loved the Chinese restaurants in New York City, and always knew where to get the best food. Actually, she was no gourmet. The restaurants 369 (in midtown), Quan Jia Fu (on Broadway), and Moon Palace (near Columbia University), were all valued ones, and she knew their owners very well. She was sad when Quan Jia Fu closed down and the owner of Moon Palace died.

Wu had her opinion of Chinese politics. She admired Deng Xiao-Ping for the "open-door policy" (leading to the recent prosperity), but disagreed with his handling of Hu Yao-Bang and Zhao Zi-Yang, setting them up as scapegoats when the situation went out of control.[63] Wu was very unhappy after the June 4, 1989 Tiananmen Square incident. She privately criticized the Chinese government for several months, and refused to see Zhou Guang-Zhao (President of the Chinese Academy of Sciences) when he visited the US in October. Her speech in the Committee of One Hundred (organized by I. M. Pei to improve communications between the American and the Chinese people) was very moving.[64]

Wu was aging gradually and stayed home most of the time after retirement. On the coffee table, there was a round marble bowl from Taiwan,

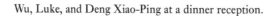

Wu met with Deng Xiao-Ping in China in the 1980s. With C. N. Yang (*second from left*), Samuel C. C. Ting (*third from left*), and Luke (*right*).

Wu, Luke, and Deng Xiao-Ping at a dinner reception.

with several red pebbles from Yu Hua Tai in Nanjing, immersed in American water. This meaningful combination might have reflected of her state of mind. Around 1960, Wu wrote to an old friend: "Someone just called saying that tomorrow is Chinese New Year. It is natural to think of hometown and family in these festive days. It is sad to think of my ailing mother, with both father and elder brother having passed away.... We are wandering in a foreign country. Even though we are established, we still have this feeling of rootlessness. I have this urge to send greetings to my old friend."[65]

Notes

1. "Small Problems Created by Great History", Huang Ren-Yu, *United Newspaper*, January 10, 11, 1994.
2. Interview with Wu Chien-Shiung, February 12, 1990, New York City residence.
3. Interview with Wu Chien-Shiung, December 5, 1989 and February 12, 1990, New York City residence.
4. Interview with Luke Yuan, June 23, 1990, New York City residence.
5. Interview with Wu Chien-Shiung, September 25, 1990, New York City residence. Interview with Luke Yuan, June 23, 1990, New York City residence.
6. *Draft of a Chronicle History of Mr. Hu Shih*, Hu Song-Ping, pp. 2570–2571.
7. Interview with Wu Chien-Shiung, June 22, 1990, New York City residence.
8. Interview with Luke Yuan, September 29, 1989, New York City residence.
9. Interview with Ren Zhi-Gong, September 26, 1989, Washington DC residence.
10. "A Historical Document — A Footnote to Our Nuclear Energy Policy", Wu Da-Yu, in *Selected Papers of Wu Da-Yu*, Vol. 6, p. 77, Yuan-Liu Press, June 1, 1989.
11. Letter to Adina Wiens from Wu Chien-Shiung, December 13, 1945.
12. *Draft of a Chronicle History of Mr. Hu Shih*, Hu Song-Ping, pp. 2861, 2862.
13. Same as 12.
14. Interview with Wu Chien-Shiung, January 6, 1990, New York City residence.
15. Letter to Zhang Qun from Hu Shih, *Draft of a Chronicle History of Mr. Hu Shih*, Hu Song-Ping, p. 2682.
16. Same as 15, p. 3892.
17. Same as 15, pp. 3898–3902.
18. There was a special profile of Wu's visit in "Taiwan Comments", February 21. *Draft of a Chronicle History of Mr. Hu Shih*, Hu Song-Ping, p. 3906.
19. *Draft of a Chronicle History of Mr. Hu Shih*, Hu Song-Ping, pp. 3916, 3917.
20. Speech given at the award ceremony for the Special Contribution Award, Chi-Tsin Culture Foundation, Wu Chien-Shiung.

21. Impression from interviews with Wu and Luke. An article in *Shin Sheng Daily News* also stated the complaints about the excessive social functions expressed by many returning scholars, July 21, 1965.

22. *Credit Newspaper*, Li Qing-Rong, July 16, 1965.

23. Interview with Wu Chien-Shiung, August 21, 1990, New York City residence.

24. Interview with Wu Chien-Shiung, September 25, 1990, New York City residence.

25. Interview with Wu Chien-Shiung, July 25, 1990, New York City residence.

26. Interview with Luke Yuan, September 29, 1989, New York City residence.

27. Interview with Luke Yuan, June 27, 1990, New York City residence.

28. *Selected Papers (1945–1980) with Commentary*, C. N. Yang, p. 77, W. H. Freeman and Company, 1983.

29. Interview with Ren Zhi-Gong, September 26, 1989, Washington DC residence.

30. *Family of Yuan Shi-Kai*, Zhou Yan, pp. 244–245, Chinese Youth Press, Beijing, 1991.

31. Interviews with Wu Chien-Shiung and Luke Yuan, May 5 and June 23, 1990, New York City residence.

32. Interview with Luke Yuan, September 27, 1990, New York City residence.

33. Interview with Wu Chien-Shiung, December 14, 1989 and May 5, 1990, New York City residence.

34. Interviews with Wu Chien-Shiung and Luke Yuan, May 5 and June 23, 1990, New York City residence.

35. Letter to Xu Jing-Yi from Wu Chien-Shiung, December 1977. Provided by Xu.

36. Interview with Wu Chien-Shiung in *New York Post*, January 22, 1959.

37. Interview with Wu Chien-Shiung in *New York Post*, October 17, 1962.

38. Interview with Wu Chien-Shiung, June 22, 1990, New York City residence.

39. Interview with Wu Chien-Shiung, July 31, 1990, New York City residence.

40. Interview with Wu Chien-Shiung in *Newsweek*, May 20, 1963.

41. Interview with Wu Chien-Shiung, September 15, 1990, New York City residence.

42. Report in *The New York Times*, November 25, 1972.

43. Interview with Ren Zhi-Gong, September 26, 1989, Washington DC residence.

44. Welcome speech to Premier Zhao Zi-Yang on his visit to the US, Wu Chien-Shiung, January 15, 1984, Plaza Hotel, New York City.

45. "Time to Go Home — Memories of Zhang Zhao-Ding", Li Yuan-Zhe, May 11, 1993, in *Reading Times*.

46. Conversation with Wu Chien-Shiung, March 1983.

47. Interview with Wu Chien-Shiung, September 13, 1989 and October 26, 1990, New York City residence.

48. Interview with Wu Chien-Shiung, June 21 and October 26, 1990, New York City residence.

49. Interview with Wu Chien-Shiung, July 25, 1990, New York City residence.

50. The assessments of her respect for the elders and care for the youngsters came from interviews with Ren Zhi-Gong and Zhang Jie-Qian. The assessments that professors

at Columbia University were not close to students came from interviews with Stanley Ruby (January 15, 1990) and Noemie Koller (December 12, 1989).

51. Interview with Wu Chien-Shiung, August 28, 1990, New York City residence.

52. The other four students were Wang Zheng-Fang, Wang Chun-Sheng, Chen Heng-Ci, and Chen Zhi-Li.

53. Telephone interview with Liu Yuan-Jun, April 28, 1995, Taipei.

54. Interview with Wu Chien-Shiung, January 27, 1990, New York City residence.

55. Interview with Wu Chien-Shiung, August 21, 1990, New York City residence.

56. Interview with Wu Chien-Shiung, October 26, 1990, New York City residence.

57. Interview with Wu Chien-Shiung, August 21, 1990, New York City residence.

58. "The Development of Sciences and Technologies in China," Wu Chien-Shiung, in *20th Annual Meeting of the Chinese Engineers and Scientists Association* (USC, February 6, 1982).

59. Interview with Wu Chien-Shiung, July 31, 1990, New York City residence.

60. Interview with Wu Chien-Shiung, April 24 and August 21, 1990, New York City residence.

61. Interview with Margaret Lewis, February 22, 1990, library at Harvard Observatory, Boston.

62. Interview with Evelyn Hu, April 1, 1990, office at UC Santa Barbara.

63. Interview with Wu Chien-Shiung, August 21, 1990, New York City residence.

64. Interview with I. M. Pei, November 29, 1989, office in New York City.

65. Letter to Xu Jing-Yi from Wu Chien-Shiung, December 1977. Provided by Xu.

Chapter 13

The First Lady of Physics Research

Why should we write yet another biography of a scientist? And why has science been so important in our culture?

The period around the turn of the 20th century was indeed a revolutionary one in the history of modern physics. The German physicist Max Planck proposed the quantum concept and the Jewish genius Albert Einstein formulated special relativity then, both establishing the foundation of modern physics which dominated the following century.

However, the reason that the 20th century was recognized as a century of science and technology was *not* these dawning concepts, or the elegant quantum mechanics which matured in the 1930s. To the general public, these physical theories were way too incomprehensive and mysterious, not unlike the old texts or chants of the ancient religions. But they were all touched by the applications and products resulting from the scientific knowledge, such as nuclear weapons, lasers, and microelectronics. The impact on human life and history was dramatic. This was the reason.

The scientists naturally became new heroes in that century. They occupied a privileged position, like high priests in the church–state period, court scholars in China, and artists during the Renaissance.

Modern physics originated in Europe. Quantum mechanics, which matured in the 1920s and 1930s, was an especially representative development. Physicists growing up at that time have done much self-examination.

Most of those physicists died near the end of the century. Victor Weisskopf became one of the living witnesses. He began as an assistant to the "great Pauli" in the 1930s.

Weisskopf wrote a book, *The Privilege of Being a Physicist*, examining the role of physicists in society. He began clearly: "Physicists enjoy certain privileges in our society."[1] He continued: "They have reasonable compensation, receive funding for laboratories and expensive, complex equipment. Unlike others, they are not expected to generate profit, but just spend the money. In addition, they do whatever they are interested in, and their only results are reports and papers, which are too difficult to be understood by the funding agencies and taxpayers anyway."

Based on the new knowledge of nature, enrichment, applications, and impact on society, scientists earn this privilege. Although Weisskopf reminded scientists of their responsibility to society, he was also firm in his belief in the importance of pure, fundamental research in science. Scientists are not totally free to do whatever they like. With massive funding, but furious competition, the scientific community has developed certain rules. The public, or even fellow scientists in other fields, might not understand the highly diverse research results, but a relative value system and standard do exist in the community.

The accumulation of new knowledge has also accelerated rapidly, helped by both massive funding and keen competition. Wu said in 1981: "Eighty percent of the scientific knowledge was accumulated in the last 30 or 40 years."[2] These new findings were valued differently, depending on the characteristics and their contribution to science and relevance to society. There are different gauges in different branches of physics, partly objective, partly with personal bias.

At the very top level, there are some absolute standards and gauges, in spite of all those opinionated physicists with big egos.

Here is an interesting story. Wu worked for Emilio Segrè, who in turn worked for Enrico Fermi. Both were Nobel laureates. Fermi once said: "Emilio, if you exchange *all* your research with a single paper by Dirac, you would still do better." Dirac was regarded as a theoretical physicist who was probably second only to Einstein in the 20th century. Segrè was not pleased with the blunt assessment, but could not argue. He shot back: "If you exchange *all* your research with a single paper by Einstein, you would still do better." Fermi thought for a minute, and agreed.[3] One can tell from this story that very few of the countless scientific papers will have lasting value. Some research made contributions, but most papers were worthless.

Science magazine discussed the "citation" issue in 1992. It stated that about 70% of papers never got cited. These papers contributed very little to the advancement of science. When visiting Taiwan in 1992, P. Griffith

(Director of the Institute for Advanced Study) also said that the "no citation" percentage would increase to 90% if "self-citation" were eliminated.

The quality of the work of a scientist definitely affects its value and impact. It is then similar to other creative works, like painting, music, literature, and sculpture.

The achievement and legacy of a scientist are determined by the assessment of the work. What is the assessment of Wu's scientific work? What are her achievement and legacy?

Most distinguished scientists have their own unique style. In Wu's 40 years of scientific research, she was most admired for the precision and completeness of all her experiments. This style of work started when she was a graduate student at Berkeley. The series of experiments on beta decay, parity nonconservation, and conserved vector current all demonstrated her demand for flawless precision.

In experimental physics, flawless precision is highly valued. Several experimental physicists, who themselves were first rate, said that Wu, as far as they knew, was the only one who had never made a mistake in her experiments.[4]

Understanding the nature of experimental physics, "not making a single mistake in lifelong experimental research" was quite an accomplishment. Even Nobel laureates had records of mistakes. A book titled *Nobel Dreams* documented many notorious errors that Carlo Rubbia (1984 Nobel laureate in Physics) had made in his experiments.[5]

The experimental physicist Maurice Goldhaber (once the Director of Brookhaven National Laboratory) assessed Wu: "We physicists believe that the most important characteristic of a great physicist is accuracy, and we absolutely value it."[6] In addition to accuracy, her other distinct characteristic was the possessing of both insight and taste in physics. With this broad insight and great taste in physics, Wu had identified truly important problems, and relentlessly pursued them with her signature precision and experimental technique. Her firm experimental evidence made history in physics. In the long history of scientific development, good scientists distinguished themselves from the rest with their penetrating insight. They had good taste, and recognized what were truly important problems. These talents are what enable cream to rise to the top.

The physicist C. N. Yang praised highly Wu's unique insight. He used the parity nonconservation experiment as an example. Wu was the very first to recognize its importance, and relentlessly conducted the experiment in the face of all kinds of hardship.[7] Other collaborators and scientists in similar research fields all praised her deep insight. Robert Serber, her colleague for many years at Columbia University, praised highly her taste in physics.[8] Her other colleague at Columbia, Charles Townes, said that her insight and knowledge were very broad, and that she was a true authority on beta decay.[9] Leon Lederman also emphasized her good taste in physics.[10]

The science reporter Walter Sullivan of *The New York Times* said that Wu gave him the impression that she was brilliant and clearly knew what the truly important problems in physics were.[11] Sullivan, who died in March of 1996, worked at *The New York Times* for over 30 years, won the prestigious Pulitzer Prize, and retired at 67 in 1987. He was highly regarded in the US scientific community, and won the first Walter Sullivan Award in Geophysical Science, established by the US Geophysical Society in his honor.

Wu's achievements can be summarized in chronological order: as a graduate student at Berkeley, she investigated the radioactive gas xenon from uranium fission. The result later solved the puzzle that chain reactions could not be maintained in the first reactor at Hanford in Washington State, and also made a critical contribution to the Manhattan Project which was developing the atomic bomb.

After the Second World War, Wu completed a series of comprehensive and precise experiments on beta ray spectra, for both allowed and forbidden transitions. She confirmed the agreement between theory and experiment, settled the dispute that had been going on for decades, and firmly established the "universal Fermi theory of beta decay". Her continuous research on beta decay established her as the world authority, and one of the prominent experimental physicists of the 1950s.

Then from the second half of 1956 to the beginning of 1957, Wu suggested and completed the parity nonconservation experiment, in collaboration with four physicists from the National Bureau of Standards. This experiment was a revolutionary development in 20th-century physics. The hypothesis of Lee and Yang was well known, but her experiment was critical in confirming its validity with very accurate evidence in record time. This established her historic legacy in physics.

Wu's next important achievement was the test of the concept of conserved vector current in 1963, put forward by Richard Feynman and

Wu, Luke, and the famous
New York Times science reporter
Walter Sullivan (*extreme right*).

Wu in front of photo of the great
physicist Werner Heisenberg.

Murray Gell-Mann (both Nobel laureates). The experiment was extremely difficult to conduct, but important. It again illustrated her talent for selecting an important problem and executing a complex experiment.

In addition to these three better-known experiments, she had worked in other areas that had demonstrated her deep insight and broad interests in physics. These additional areas included double beta decay and lepton conservation, a test of the "Einstein-Podolsky-Rosen argument" in quantum mechanics, exotic atoms, and the Mössbauer effect and its applications in biochemistry. In device physics, she achieved much in high-sensitivity radiation detectors, and a large-scale, low-temperature environment.

Although theories and experiments are interrelated in physics, theorists and experimentalists have somewhat different perspectives. Experimental physicists firmly believe that physics is an *experimental science* and the experimental evidence has the final say. Wu belonged to this school.

People have the impression that great theorists are brilliant, with talents shining like comets in the sky — blindingly bright. They generally had their best work done in their 20s. The four most important figures in quantum mechanics — Dirac, Heisenberg, Pauli, and Bohr — all did their great theoretical work at a young age. Indeed, Dirac and Heisenberg were both escorted by their mothers to Sweden to receive their Nobel Prizes.

Dirac wrote a poem referring to "30 years old as a limit". He said: "Age is a slowing factor, which every physicist should be aware of. Once the 30th birthday is passed, he is better dead than alive."

On the other hand, Wu, like many other experimentalists, had a long creative career. The 1971 article "Wu Chien-Shiung, The First Lady of Physics Research" said: "Wu Chien-Shiung, at 57, is still a leader in experimental physics. Most physicists withdrew from active competition in their 40s."[12] Her long, lasting achievements earned her the admiration and respect of physicists worldwide. But something was still missing — the Nobel Prize.

Without the Nobel, she did not receive the proper recognition from the public. Many scientists continue to resent this injustice.

When Lee and Yang won the Nobel Prize for parity nonconservation, Robert Oppenheimer, Wu's professor at Berkeley, said in public that she should also be honored. He pointed out that Wu and Emilio Segrè were probably the only two physicists in the world capable of doing the experiment on parity violated beta decay.[13] Both Lee and Yang believed that Wu should win the Nobel Prize.[14] Yang had since nominated her several times.[15]

Isidor Rabi, who single-handedly developed Columbia's physics department into a premier institution, said at a public meeting in 1986 that Wu deserved to win the Nobel Prize.[16] Robert Wilson, Founding Director of the Fermi National Laboratory, said the same thing.[17]

Herwig Schopper, who once served as the Director of CERN, strongly believed that Wu should have received the Prize together with Lee and Yang, as she was the first to confirm parity nonconservation experimentally.[18] Jack Steinberger, her colleague at Columbia, wrote an article in *Science* expressing his view that Wu should receive the Nobel Prize.[19] He also believed that the omission of Wu when Lee and Yang won the Nobel Prize was the single worst mistake that the Royal Swedish Academy had ever made.[20]

Norman Ramsey and Glenn Seaborg, both Nobel laureates, believed that Wu deserved the Nobel Prize without any question.[21] James Rainwater, another of her colleagues at Columbia in the 1940s and 1950s, said that he was thunderstruck when he received the news that he had won the Nobel Prize in 1975. He called Wu and asked her to represent him in receiving the Nobel Prize, believing that she was more worthy of the honor.[22] Nicholas Samios, once the Director of the Brookhaven National Laboratory, said interestingly: "I always thought that she had won the Nobel Prize. She belonged to this community of laureates in my mind."[23]

In 1980, Nobel Prizes were awarded to J. Cronin and V. Fitch for their experimental evidence that both charge conjugation and parity conservation were violated. Many scientists believed that Wu should have been included that year.[24]

The Nobel Prize has its supreme prestige in science partly because of its rigorous selection process. It has been awarded annually starting in 1910, after the nomination of candidates in physics and chemistry by the domestic (about 300) and foreign (about 160) members of the Royal Academy in Sweden. Nomination also comes from past Nobel laureates, the Nobel Committee (five members each in physics and chemistry), full professors in Sweden, Denmark, Finland, Iceland, and Norway, at least five chair professors in the world selected by the Royal Academy, prominent professors, directors of research institutes, and distinguished scientists invited by the Royal Academy.

The Royal Academy solicits nomination from the above 2000 to 3000 people each year (about 10% response), and with a deadline of January 31. The selection committees recommend no more than three winners in a field by the end of September, with the winners approved by the Royal Academy and announced in the middle of October. The prize money was 150,000 Swedish kronor in 1901, and has increased to 6,700,000 since 1993. Although the Nobel Prize involves a rigorous selection process, it can still be influenced by personal factors. There have been disputes. In 1995, there were rumors that a pharmaceutical company had heavily promoted a winner in medical science a couple of years earlier. In addition, some biographies document the marketing of a candidate in the nomination process.

That Wu never won the Nobel Prize might indicate that some scientists did not think that she was worthy of the prize. Her series of experiments on beta decay were important achievements, but not a "major invention or discovery" as stated in the original Nobel criteria. Others might argue that she did not initiate the experiments on parity nonconservation and conserved vector current, as both were suggested by theorists. One other reason, impossible to verify, was that Wu had made an enemy of one scientist close to the Royal Academy in Sweden.[25]

Her reputation did not suffer for not winning the Nobel Prize. She was admired by physicists in Europe and the US, even by scientists in Russia. The Russian genius theorist Lev Landau and A. I. Alikhanov (who founded the Institute of Theoretical and Experimental Physics at the Russian Academy) praised highly her achievements. Alikhanov once simply referred to parity nonconservation as "Wu's effect" in China.[26]

Wu received many awards. The physicist Pu Da-Bang invited her to visit Taiwan in 1983. Introducing her at the National Central University, he said: "We can say that Wu has received all the important prizes and awards except for the Nobel." Her first honor came when she was elected a fellow of the American Physical Society in 1948 at the age of 36, when she had already done much in the field of beta decay. Wu was elected to the prestigious National Academy of Sciences in 1958 — the seventh woman member. On June 17 of the same year, she was awarded an honorary Sc.D. by Princeton University — the first woman in its 211-year history. Lee and Yang were awarded that honorary degree at the same time. All three were elected to Academia Sinica. On January 22, 1959, Wu was the first woman to receive the Research Corporation Award, for her contributions to the field of beta decay and the parity nonconservation experiment, in New York City.

She received the Achievement Award from the American Association of University Women on June 23, 1958, and an honorary Sc.D. from Smith College in the same year. The next year, she received an honorary Sc.D. from Goucher College.

On October 16, 1962, Wu received the John Price Wetherill Medal from the Franklin Institute in Philadelphia for the parity nonconservation experiment. Her collaborators Ernest Ambler, Ralph P. Hudson, Dale D. Hoppes, and Raymond W. Hayward were also given medals. Two days later, she received the Woman of the Year Award from the American Association of University Women.

In 1963, she received an honorary Sc.D. from Rutgers and was invited to be a member of the Advisory Board on Natural Sciences of the Chinese University of Hong Kong.

Wu was given the Cyrus B. Comstock Award of the National Academy of Sciences on April 27, 1964, at their 101st congress in Washington DC. The President, Frederic Seitz, later became the first foreign science advisor to Taiwan in the 1980s. The Cyrus B. Comstock Award was bestowed once every five years to the very top scientists. Wu was the 11th recipient, and the first woman. Of the ten previous winners, six won the Nobel Prize.[27]

Wu returned to Taiwan in 1965 to collect the first Achievement Award from the Chi-Tsin Culture Foundation. On June 12, 1967, she received an honorary Sc.D. from Yale University; one other recipient was the world-renowned jazz musician Duke Ellington. In 1969, she was elected an Honorary Fellow of the Royal Society of Edinburgh. On May 12, she received an Honorary LL.D. from the Chinese University of Hong Kong — the first woman recipient.

In 1971, Wu received an honorary Sc.D. from Russell Sage College, and the Golden Anniversary Award from the Sigma Delta Epsilon Society.

In October of 1971, *The New York Times Magazine* published a list of distinguished Chinese, including Wu, C. N. Yang, I. M. Pei, T. D. Lee, Gu Wei-Jun, and Dong Hao-Yun. Wu was invited to give the Morris Loeb

Lecture in physics at Harvard University in 1973–1974. In 1974, she won the Scientist of the Year Award of *Industrial Research Magazine*, a top award from the US industry. She also received an Honorary Sc.D. from Harvard University, Bard College, and Adelphi University. She received the Boris Pregel Award from the New York Academy of Sciences for research in nuclear physics and nuclear engineering.

In 1975, Wu received the Tom Bonner Prize for nuclear physics from the American Physical Society, and an honorary Sc.D. from Dickinson College. She also received the National Medal of Science, the very top national honor, from US President Gerald Ford on October 18, 1976 in the White House.

In 1978, an Israel industrialist established the Wolf Prize, to specifically honor the most distinguished scientists in the world who were overlooked by the Nobel Committee for whatever reason. Wu was awarded the first Prize in Physics.

Wu was inducted into the 100 Fellows Club of her alma mater UC Berkeley. She received an honorary Sc.D. from the University of Pennsylvania on May 19, 1980. The following year, she won the Woman of the Year Award of the St. Vincent Culture Foundation, sponsored by UNESCO and given by the President of Italy.

In 1981, the American Physical Society published a calendar to celebrate its 50th anniversary. It used photos of Wu, Einstein, Fermi, the Hungarian Leo Szilard, the Indian S. Chandrasekhar, the Austrian Victor Weisskopf, the Pole Maria Mayer, and others as distinguished immigrants who had contributed to the great success of US physics research.

Wu received three honorary Sc.D. degrees in 1982, from Columbia University, the University of Southern California, and the State University of New York in Albany in the US. Three honorary professorships were bestowed on her by Nanjing University, the University of Science and Technology of China, and Peking University (the oldest university in China).

In 1983, Wu was invited by the Japanese Physical Society to deliver the Nishina Memorial Lecture at the universities of Tokyo, Osaka, and Kyoto. She also received the Lifetime Achievement Award from Radcliff College, Harvard University. The other eight recipients included the artist Georgia O'Keeffe.

Wu chatting with attendees after she received an honorary doctorate from Adelphi University in 1974.

Wu received the Golden Plate Award from the American Academy of Achievement and the Women's Achievement Award from the New York City Municipal Government in 1984. She also received an honorary Sc.D. from the historical Padua University in Italy, presented in the auditorium where the father of science, Galileo, had lectured.

She collected the Second Blue Cloud Award, for outstanding contributions to culture exchange between China and America, from the Institute of China in 1985. The other recipient was the composer Zhou Wen-Zhong, a professor of music at Columbia University. In 1986, The Mayor of New York, Edward Koch, awarded her the second Mayor's Award of Honor for contributions to science and technology.

Wu received the Ellis Island Medal on October 27, 1986. This award was established to celebrate the Centennial of the Statue of Liberty on Ellis Island. Fifteen thousand distinguished immigrants were nominated, and Wu

232

Wu was awarded an honorary doctorate by the President of Columbia University, Michael Sovern, in 1982.

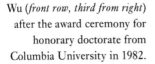

Wu (*front row, third from right*) after the award ceremony for honorary doctorate from Columbia University in 1982.

Wu was given the Mayor's Award of Honor for Contribution to Science and Technology by the New York City mayor, Edward Koch (*extreme right*), in 1986.

With boxing champion
Muhammad Ali after
receiving the Ellis
Island Award in 1986.

With Isidor I. Rabi after being awarded an honorary
doctorate by Columbia University in 1982.

was one of the 80 recipients, including the architect I. M. Pei, Zbigniew Brzezinski (National Security Advisor to President Jimmy Carter), and the boxing champion Muhammad Ali. Wu and Ali took a picture together at the award ceremory.

The Association for Women in Science published a calendar in 1988, with pictures of 13 distinguished women scientists. Six of them, including Wu, deserved a Nobel but were passed.[28] Wu received an honorary Sc.D. from her alma mater the National Central University in 1989. The Purple Mountain Observatory of the Chinese Academy of Sciences in Nanjing named International Asteroid #2752 the "Wu Chien-Shiung Star" in 1990. Columbia University awarded Wu the Pupin Medal for outstanding services to the nation in science on November 19, 1991.

In 1994, Wu together with Luke received the Achievement Award of the Institute of China. Wu, C. N. Yang, Samuel C. C. Ting, and others were elected the first foreign academicians of the Chinese Academy of Sciences. On November 6, Wu won the Ettore Majorana (Science for Peace Prize).

In addition to all those prizes and awards, Wu's career had been properly established. Columbia University initiated the title of Pupin Professor of Physics in 1973, and named Wu as the first recipient. Michael Pupin made lasting contributions to US science as well as Columbia University in the early 20th century.

Wu's achievements earned her many titles. In the citation accompanying the honorary Sc.D. from Princeton University, she was honored as the "top woman experimental physicist in the world". She was later honored as the "queen of nuclear physics".[29] When appointed as the first Pupin Professor, she was affectionately called "the first lady of physics research".[30] Wu was also praised as "a great teacher and effective educator for the young generation", as well as "warm and honest; her achievements crossed national and cultural boundaries".[31]

When naming her the first Pupin Professor, the public relations department of Columbia University prepared for a press release and asked Wu for her input. She was too modest to talk about herself; rather, she took the opportunity to highlight the achievements of Pupin. In particular, she discussed the great invention of the Pupin circuit, which was later critical in the transmission of long distance telephone communications. Wu said

In Tokyo University, the calligraphic
piece "Mental Examination Based
on Systematic Observations" and its
view of the universe impressed Wu
profoundly.

In the laboratory of Professor Nishina of Tokyo University in 1977.

that the story of Pupin, from an immigrant to a great inventor, was even unknown to people walking in and out of Pupin Laboratories daily.

Wu remarked that Pupin had great love for fundamental science. US science was rather backward at the turn of the 20th century. Pupin diligently promoted its importance, and established both the American Physical Society and the American Mathematical Society at Columbia University.[32] The senior Pupin leveraged his reputation to advance fundamental science in the US. He promoted idealism in science, lecturing at many universities and institutes. Freeman Dyson of Princeton said in an article that Pupin, as a pragmatic inventor, contributed much more than his peers to the idealism in both science and US society. Wu assessed Pupin: "His work always demonstrated a profound insight, and a great combination of theoretical deduction and experimental execution." Wu's colleagues agreed that the same appraisal would apply to her.[33]

Notes

1. *The Privilege of Being a Physicist*, V. Weisskopf, W. H. Freeman & Company, New York, 1989.
2. English lecture at the Sixth Annual Meeting of the Institute of China, Wu Chien-Shiung, November 14, 1981.
3. *A Mind Always in Motion: The Autobiography of Emilio Segrè*, Emilio Segrè, UC Berkeley Press, 1993.
4. These physicists included H. Schopper, a one-time director of CERN, and V. Telegdi, a Wolf Prize recipient in 1990.
5. *Nobel Dreams*, Gary Taubes, Random House, New York, 1986.
6. Interview with Maurice Goldhaber, April 10, 1990, Brookhaven National Laboratory.
7. Interview with C. N. Yang, November 16, 1989, New York City.
8. Interview with R. Serber, March 14, 1990, New York City residence.
9. Interview with C. Townes, March 29, 1990, office in UC Berkeley.
10. Interview with L. Lederman, March 27, 1990, office in the University of Chicago.
11. Interview with Walter Sullivan, December 15, 1989, Korean restaurant in New York City.
12. *Smithsonian Magazine*, p. 52, January 1977.
13. Interview with Wu Chien-Shiung, February 12, 1990, New York City residence.
14. Interview with C. N. Yang, September 12, 1989, office at SUNY Stony Brook; Interview with T. D. Lee, October 2, 1989, office at Columbia University.
15. Interview with C. N. Yang, December 16, 1989, Doral Inn Hotel in New York City.

16. *New York Times*, speech given at the Annual Award Meeting of the US Nobel Celebration, I. I. Rabi, December 11, 1989.
17. Interview with R. Wilson, September 21, 1989, Ithaca, New York City residence.
18. Interview with H. Schopper, August 6, 1989, office at CERN.
19. Science, p. 669, November 4, 1988.
20. Interview with J. Steinberger, August 9, 1989, office at CERN.
21. Interview with N. Ramsey, February 22, 1990, office at Harvard University.
     Interview with G. Seaborg, March 28, 1990, office at Lawrence Berkeley Laboratory.
22. Interview with Wu Chien-Shiung, April 5, 1990, New York City residence.
23. Interview with N. Samios, February 26, 1990, office at Brookhaven National Laboratory.
24. Many great scientists, including C. N. Yang, held this opinion.
25. This speculation was based on private conversations with Wu and many other physicists. It is reasonable, but impossible to verify.
26. Interview with Wu Chien-Shiung, December 14, 1989, New York City residence.
27. The ten recipients including six Nobel laureates (and the years) are Robert Millikan (1932), Samuel Barnett, William Duane, C. Davidson (1937), Percy Bridgman (1946), Ernest Lawrence (1939), Donald Kerst, Merle Tuve, William Shockley (1956), and Charles Townes (1964).
28. The 13 distinguished scientists (and the year of the Nobel Prize) were: Madame Curie (1903, 1911), Irene Curie (1935), Gerty T. Radnitz Cori (1947, Medicine), Maria G. Mayer (1963), Dorothy C. Hodgkin (1964, Chemistry), Rosalyn S. Yalow (1977, Medicine), Barbara McClintock (1983, Medicine), and Rita Levi-Montalcini (1986, Medicine); the others were German mathematician Amalie Noether (should have shared with Einstein the 1921 Prize), German embryologist Hilda P. Mangold (should have shared with H. Spemann the 1935 Nobel), Lise Meitner (should have shared with O. Hahn the 1944 Nobel), Wu (should have shared with Lee and Yang the 1957 Nobel), Rosalind E. Franklin (should have shared with Watson and Crick the 1962 Nobel), and Jocelyn Bell Burnell (should have shared with M. Ryle and A. Hewish the 1974 Nobel).
29. Emilio Segrè called her this name. *Newsweek*, May 20, 1963.
30. Press release on the appointment of Wu as the first Pupin Professor, Columbia University, January 5, 1973.
31. Same as 30.
32. Letter from Wu to the public relations department of Columbia University, January 1973.
33. Press release on the appointment of Wu as the first Pupin Professor, Columbia University, January 5, 1973.

Chapter 14

Wu Chien-Shiung the Scientist

In February 1978, Wu Chien-Shiung received a letter from the Wolf Foundation of Israel announcing her selection as the first recipient of the Wolf Prize in Physics. The award ceremony was to be held in the Israeli capital on April 10.

Wu had never heard of the Wolf Prize, and did not read the letter carefully. The prize money was actually comparable to that for the Nobel Prize, around US$100,000.[1] But in the letter, the prize money was written as US$100.000 (European convention). Wu joked with some colleagues and students that she would not go all the way to Israel to receive just 100 dollars.[2] Her secretary noticed the difference. There were five Wolf Prizes to be awarded each year, for physics, chemistry, mathematics, agriculture, and medicine. Each prize had a selection committee of three international professors. In the prize's first year of existence, four of the 15 committee members were Nobel laureates. One of the principles behind the Wolf Prize is to honor distinguished scientists who should win the Nobel Prize but have been overlooked. As it has very stringent selection criteria, and its selection committee members were internationally renowned, it gained international prestige rapidly.

Wu had never been very interested in, and had never paid much attention to the awards. She never fully enjoyed the fame that accompanied her accomplishments, especially after the parity nonconservation experiment.

Once, she was asked to approve the release of an article with the headline "Dr. Wu, born in China, the top woman physicist in the world", and she felt very annoyed. She wrote to the public relations department of Columbia University: "I do not enjoy exposure, and do not like shocking headlines".[3] Wu had always been a quiet, solid researcher, and disliked flashy distraction. Her collaborators and students remembered her as

Wu was awarded the (first) Wolf Prize of Israel in 1978. With Prime Minister Menachem Begin on her right.

being very focused on her work, not noticing any other thing around her. She liked to work in her own laboratories, an environment where she had total control.[4] Her manner was a reflection of her personality. She was basically a private person, and was rather reserved. Some felt that they did not know what her true intentions were.[5]

Wu was also contemplative and quiet. Her friend Zhang Xi-Ying from the National Central University days said that Wu enjoyed small gatherings with a few old friends, but became awkward in a big group, not knowing what to say.[6] Wu would fall into deep thought in the middle of a conversation, or suddenly jump to another topic. On a couple of occassions, Luke tried to make her aware of her behavior, worrying that she might offend others. He said that she had been more restrained as a maiden but had become more comfortable with her own way.[7]

Even though Wu was private, sometimes opinionated, and not interested in self-promotion, many scientists enjoyed their collaborations with her, and had benefited much from her insight into science.[8] Wu was loyal to her friends and straightforward with them. She was considerate, always bringing along small gifts when visiting friends abroad. Her foreign friends typically refered to her as "decent".[9] People had different opinions as to whether she had a sense of humor, depending on how close they were to her. Her old friends, especially nonscientists, thought that she was humorous and would sometimes laugh at her own words.[10]

The physicist Jack Ullman, who had collaborated with Wu on the double beta decay experiments in the 1960s, had a personal experience showing that Wu was a very serious person.

Ullman and Wu once discussed ways to become a great experimental physicist. Wu said that experimentalists should be reasonably clever, but it was not the *most* important factor, as it was in the case of theorists. The most important factors were persistence, good decision-making, and a little luck. When Ullman recounted this discussion to a woman librarian, she joked that Wu was subtly telling him that he was *not* clever enough. They both thought that it was funny. Encouraged by the librarian, Ullman told Wu about this joke the next day. Very seriously, Wu said: "Why did she say so? You *are* very clever."[11]

Wu closely kept track of major happenings in society, politics, and the economy. Time permitting, she read in detail (and formed her opinions) *The New York Times* and other newspapers, as well as magazines. But she was quite ignorant about other entertaining topics. She once overheard a discussion by graduate students describing their party as an "orgy", but had no idea what it meant.[12]

Her manners were basically very traditional Chinese ones. She demanded a lot from students, but also cared for them a lot. Her student Noamie Koller recalled that her laboratory work was particularly busy when she got married. Wu gave her time off for her honeymoon — starting on a Thursday after an examination, but coming back to work on the following Monday. Wu also gave her the biographies of E. Fermi and R. A. Millikan to read on the honeymoon.[13]

Wu imposed the same kind of demand on herself. When she went on vacation to the Bahamas with Luke, she brought along the book *The Double Helix: A Personal Account of the Discovery of the Structure of DNA* (by the Nobel laureate J. Watson) to read. She joked that it was the only day when she did not touch physics.[14]

Her many students recalled that her caring attitude toward students was highly unusual in the US and Europe. Wu cared about their friends, marriages, families, and careers, and helped out when possible.[15]

She was also quite strict about maintaining the traditional teacher-student relationship. She was properly respectful of her teachers Hu Shih and Gu Jing-Wei, and privately scorned young scientists who disowned their teachers once they became established.[16] At one party given by Wu at her home for her students and colleagues, a student experienced a slip of the tongue, calling her Gee Gee, her nickname at UC Berkeley. She was displeased and told him: "There is only one person here senior enough to call me Gee Gee." The person was Robert Serber, who went to UC Berkeley before her, and had been her colleague at Columbia University for many years.[17]

She benefited from having a very supportive and considerate husband. Her friends Adina Wiens and Margaret Lewis both approved of Luke, thinking that Wu had made the right choice. When Adina met Luke for the first time at Lake Tahoe in 1941, she declared: "Gee Gee, he is your man."[18] Many of their friends believed that Luke was so complementary and supportive that he deserved a lot of credit for her success. At the peak of her career, a busy and confident Wu could be bossy and opinionated. Luke took it well, remaining considerate, and often yielding to her way. Their friends admired his even temper and tolerance.[19]

Being both physicists but in different areas of physics, they worked and lived in different locations at times. Luke was in high-energy physics, and had once worked at CERN and in France. He worked at the Brookhaven National Laboratory for many years, living in Long Island on weekdays and spending weekends with Wu in Manhattan. They attended separate conferences in their own areas.

Wu and Luke did collaborate. In 1961 and 1963, they authored the book *Methods of Experimental Physics*, Volumes 1 and 2, together. Later, Luke wrote the book *Nature of Matter: Purposes of High Energy Physics* in 1964. Wu wrote the classic *Beta Decay* together with theoretical physicist S. A. Moszkowski in 1966.

Luke was not as accomplished as Wu, but he did make contributions in a number of areas. In his early cosmic ray experiments, he demonstrated that neutrons must be secondary particles. He worked on high-frequency

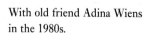

With old friends Adina Wiens
(*right*) and Eda, (*left*) in the 1970s.

With old friend Adina Wiens
in the 1980s.

systems in the high-energy accelerator at the Brookhaven National Laboratory, and discovered the baryonic resonance states in high-energy scattering experiments. His most important contribution came from his research on transition radiation.

A Russian theoretical physicist first proposed transition radiation. The radiation emitted by a very-high-energy particle as it penetrates various materials was subsequently observed in accelerators, and generated much interest. With his early background in the field of cosmic rays, and later in high-energy experiments, Luke investigated this phenomenon and published numerous results.

In 1972, he sent a paper of his research on transition radiation to Luis W. Alvarez, who won the Nobel Prize in Physics in 1968. With his talent, creativity and invention in several fields, Alvarez was known for his big ego. He replied near the end of the year, saying that he had now been convinced of the importance of transition radiation after reading Luke's paper, and he praised highly his research work.[20]

Luke maintained his interest in transition radiation, but did not make much progress, as the detection of such a minute amount of energy was too difficult. When a French research group considered using a low temperature superconducting detector, Luke applied for funding support from the Defense Department of Taiwan in 1994 to further pursue this effort.

Luke was soft-spoken, careful, and patient. The couple's friends all admired his gentleness.[21] He was a dedicated, hands-on scientist. In daily life, he was interested in fixing appliances around the house, and was a good cook to boot.

He had broad interests in music and photography. As a student at Yenching University, he learned to play the *nan hu*, taking lessons from the master musician Liu Tian-Hua. He was quite skillful, and he even performed on stage once in Beijing. But the story that he was a student of Cheng Yan-Qiu was not true. As Luke was the grandson of Yuan Shih-Kai, the last emperor, he was in an elite social circle. He had a young uncle who had known Cheng Yan-Qiu very well. He and the uncle had dinner one day, and met Cheng dining in the same restaurant. All of them took an after-dinner cruise on Bei Hai Lake. Cheng was in a good mood and decided to sing a couple of arias, and Luke followed with another aria. People in the other boats enjoyed the singing, and believed that the singers belonged to the same school. So the rumor that Luke was a student of Cheng started to cir-culate.[22] Luke occasionally practiced the *nan hu*, and performed a couple of times at UC Berkeley. He then gave it up, as the practice was too time-consuming, but sometimes he still performed a couple of pieces to entertain guests.[23]

Many physicists knew about the special background of Luke, that his grandfather was the last emperor of China. On one occasion, Norman Ramsey, Murray Gell-Mann, and others went to a Chinese restaurant after a committee meeting in Washington DC, and Ramsey recognized the por-trait on the wall as that of Yuan Shih-Kai.[24]

Luke rarely mentioned his grandfather, or his blue blood. This was partly due to the pressure being in a prominent family, and partly because he hardly knew his grandfather. As it happened, the father and uncle of Wu actually revolted against Yuan Shih-Kai when he declared himself emperor. Wu teased Luke about this coincidence.[25]

Wu had a singular focus on science. She occasionally read books, but otherwise had no other interests. There were paintings and calligraphic pieces by famous artists, mostly gifts, in their home. She liked them, but did not particularly appreciate them. She did not enjoy traveling, and was basically a quiet, contemplative lady. One of her dear friends recalled seeing a Broadway show with her. Wu fell asleep midway through, and was awak-ened by the applause at the end.[26]

Other friends and students also said that Wu did not enjoy entertainment events or performances, and her readings in literature, drama, and music were rather limited.[27] Once, a classmate of the UC Berkeley days visited New York, and invited Wu to see the hot Broadway show *Cats*. Not knowing that the show was very popular and its tickets were hard to get, Wu did not show up, wasting all the effort made by her friend.[28] Wu herself said that she liked plays with messages, and did not care very much for musicals.[29]

Before her retirement, Wu was always busy with research, teaching and related work in the laboratory, and reading and editing at home in the eve-nings. She cared very much about the welfare and education of her son Vincent, and she occasionally cooked. Because of his mother's busy sched-ule, Vincent learned to take care of himself, and he became independent and rather quiet. With a successful and famous mother came a certain pres-sure on him. He once told Wu that it was really not easy to be her son.[30]

Wu herself did not appreciate the pressure, thinking that all would be fine as long as mother and son were not working in the same institute. She insisted that he attend the well-known Bronx High School of Science, although he was reluctant to leave his friends in Manhattan. He did transfer to the "star school", and appreciated her decision many years later.[31]

When Vincent graduated from high school, Wu and Luke asked C. N. Yang to advise him on his career. As both of his parents were in physics, Yang suggested that Vincent should consider biological science, perhaps at Yale or another university away from New York City.[32] Vincent ended up going to Columbia University and majored in Physics. He said that many of his friends were in New York City, and he loved the activities that the city offered. He was an undergraduate from 1963 to 1967, and stayed in the graduate school. He worked with Leon Lederman for his Ph.D. research.

Vincent spent a fairly long time at Columbia. There were lots of antiwar movements (Vietnam War) on campus in the late 1960s and the protesting students once occupied the administration building. Vincent was quite active, growing long hair and all that. Wu and Luke were privately worried. There was perhaps more respect than intimacy in the mother-son relationship. Vincent said that his parents never forced him to do anything.[33] But Wu once accidentally revealed that Vincent did not quite understand her,[34] and she did not think that an interview with him would be useful.[35]

Lederman described Vincent as "super cautious". After a postdoctoral fellowship at the University of Illinois, Vincent began working at the Los Alamos National Laboratory. He gained a better appreciation of Wu's accomplishments through the literature. Vincent was married to Lucy Lyon in 1974, and his daughter Jada was born in 1978. Wu just adored this granddaughter. Wu was always meticulous and elegant-looking. So one of her friends was totally surprised when told that Wu and Jada were crawling around on the floor playing with each other.[37]

Wu was modest and proper, unwilling to reveal her private life. She was very different from the colorful physicist Richard Feynman, who wrote a book telling of his many adventures. She admired Feynman as a distinguished physicist, but did not care too much for his flashy personality.[38] There were several encounters between Wu and Feynman. Feynman visited her laboratory in 1957 to find out the latest news on parity non-conservation.

Wu also confirmed the conserved vector current hypothesis proposed by Feynman and Murray Gell-Mann.

As a technical advisor to the US State Department, Wu attended the Second International Conference on the Peaceful Uses of Atomic Energy in Geneva in 1958. Feynman was there too. When Wu asked where she could reach him in case there was a message, Feynman was embarrassed as he was staying in a small hotel in the red-light district in Geneva.[39]

Feynman had to deliver a lecture, with a lot of material but not enough time. He asked Wu for advice. She simply suggested omitting a portion — nobody would notice it anyway.[40]

This was the typical Wu. She had penetrating insight into many things, but was unwilling to reveal her opinion just to get attention. This personality also drove her assessment standards. Among the many talented people, her most-favored ones might not have been the famous ones. For example, she admired the physicist E. Purcell of Harvard University, calling him a "saint".[41] Huang Hui, General Manager of Taiwan Electricity, was another one she admired.[42]

She was very selective about and loyal to friends, and ready to offer unsolicited advice. She treated the late artist and her classmate at the National Central University Sun Duo-Ci as if she was her own sister. Wu advised Sun on her romance with her teacher, the great artist Xu Bei-Hong, and took good care of her in 1954 when she visited New York City for her art exhibition.

Sun wrote an article recalling that they were thrilled at their reunion in New York. Wu said: "I know you are diligent, but you are sometimes willful and risky. I wish you would stay in the US for a couple of years, and go to Europe for some time. You will learn a lot more than staying in China." Wu arranged to have Sun do some research work at Columbia University, and asked her to move into their apartment. Sun had to prepare for her art exhibition, and was afraid of interfering with Wu's research. She did not move in.

Once her art exhibition in the gallery on 57th Street was over, Wu suddenly showed up in her hotel room and forced her to move. Sun wrote, "It was hot. I had been staying in a missionary hostel for women. The windowless, small room was inexpensive. Wu said she could not let me stay in this miserable place, and moved me into a large bedroom with a private bath in her apartment. She introduced me to an old woman professor at Columbia University to learn French, and another two Fine Arts professors there. I suddenly felt at home, and enjoyed being pampered by a dear sister. I really did not want to leave the US, if not because of missing my children."[43]

Wu frequently attended activities of the alumni association of the National Central University. She would update the status of her friends, and entered it in the yearbook.[44] In a letter to her friend, she wrote: "How is Brian? You should go and see him. He needs you, without saying so." Brian was the son of her friend.[45]

Wu was invited to deliver a lecture at the Centennial Celebration of the American University in Beirut, Lebanon in April 1967. The other two US speakers were Nathan M. Percy, President of Harvard University, and the chemist James Conant.

Her dear friend from the UC Berkeley days, the physicist Salwa Nassar, was Lebanese. Nassar went back to Lebanon after graduation, and became the President of Beirut Women's College. She strongly urged Wu to accept the invitation in 1965.[46] Wu was all excited about meeting Nassar after some 20 years. Unfortunately, Nassar suddenly died on February 16. Her student visited Wu several times, but could not gather the courage to tell her the sad news until a week before her departure. Wu was devastated.[47]

She was also very weak from an infection in March, and Luke urged her to cancel such a long trip. Wu said that she could not cancel the trip at the last minute. She left on April 17, but yielded to Luke's advice and stayed overnight in London.

Wu received a very warm welcome in Beirut. Both the English and the Arabic newspapers reported her visit in detail, praised her achievements and lecture, and voiced the belief that her visit had brought much respect and pride to modern Lebanese women. What impressed Wu most were the hospitality of Nassar's family and friends, the beauty of her country, as well as her fresh grave on top of a small hill.[48] She wrote a five-page letter to her friends from the UC Berkeley days during her long flight back to the US. She finished it saying: "I know that I have to return to work as soon as I get home. The letter that I promised you in the postcards I sent out in Beirut will never get done."[49]

The most moving part of the letter was her account of her visit to Nassar's grave.

Wu, accompanied by Nassar's sister and father, went to the gravesite on the hilltop on a Friday morning. She wrote in the letter: "When we arrived, the cloud gathered. It was haunting, and we could not hold back our tears." The letter was a rare revelation of her emotion, and illustrated her elegant writing.[50] When she first arrived in Berkeley, Wu's English was just adequate. With serious study and practice, she had mastered the language in her later days.

Wu also cared strongly about social issues. Hers was not an image of scientist living in an ivory tower, isolated from the real world. She followed world events, reading *The New York Times* and watching Public Broadcasting System TV. Her interest in special issues was enough for her to investigate them and she formed her own opinion.

Because of her busy schedule, she could not take part in many activities. She was particularly active in 1975 as the President of the American Physical Society. She greatly improved communications with the media in order to popularize science, and wrote letters to the Russian and Chilean governments expressing concern over the human rights of scientists. She and Luke took part in the march at the United Nations protesting the massacre in Cambodia.

In the 1970s, Wu used the Mössbauer effect in nuclear physics to study the structure of hemoglobin, which if defective causes sickle cell disease. It was a prime example of scientists contributing to the management of social problems. Sickle cell decease is a genetic decease that affects many blacks. The Harlem district at the edge of Columbia University had a large black population. Wu said: "Living and working near Harlem gave us a first-hand understanding of patients suffering from sickle cell disease. They are miserable."[51]

The defective hemoglobin has a half moon shape and causes blockage in small blood vessels. The patients are in pain from such blockage. Wu led a group of scientists and collaborated with physicians at St. Luke's Hospital. Their research produced some results. A biochemist reported in 1971 that they had successfully treated 25 cases.

Wu herself believed that "this case illustrated that even a seemingly remote, fundamental nuclear research technique can benefit society".[52]

On the human rights issue, she had an interesting thought. She felt that life was short and therefore everyone should live fully, with human rights. No one should get short-changed, and hope for better treatment in the next life.[53]

Wu bemoaned the passing of dear friends as she aged. She had felt immortal when young. Watching the passing of once powerful figures like Winston Churchill, Dwight D. Eisenhower, and J. Robert Oppenheimer made her realize that death was just an unavoidable natural process.[54] She watched trees in friends' gardens; trees would die when the time came. She said that even medical doctors could not live forever. God dictated the timing, and doctors might help.[55] However, she was never religious. Her friends tried to convert her and she did not protest against their good intentions, but she never sought peace of mind in religion.[56]

With the rapid passing of time, Wu believed in hard work when still young and vigorous. During an interview, she said: "I am 80 years old this year. Eighty years passed just like that. One was vigorous and productive in middle age. He was just a child before that, and became a retiree with less energy and the associated old age problems afterward. One should grasp the opportunity in middle age. There is not much time left."[57]

Wu liked to have photographs taken when she was young and pretty.[58] She had trouble accepting the fading youth in middle age. One of her female students said that Wu had two different ages at a time, but somehow this got resolved.[59] Wu felt old in the 1990s, complaining how her memory was fading.[60]

Wu was born on May 31, 1912. In her youth she never celebrated her birthday believing in the Chinese tradition that you celebrate only when you have accomplished something in life. This was the same thing her father had practiced.[61] Wu taught at Columbia University for 36 years, and received many prizes and awards in addition to her salary. She also bought stocks of blue-chip companies for long-term investment, and managed to save a fair amount.

She lived modestly. She donated about one million US dollars of her savings to establish the Wu Zong-Yi Memorial Foundation at the Ming De School in Liuhe. Her father, Wu Zong-Yi, founded the school. The annual interest income of the foundation would be used to fund scholarships for the students and provide training grants for the teachers, as well as for construction and purchase of computers.

Wu had internal discipline. She seldom openly criticized others. Her public speeches showed that she was optimistic and had firm belief in science. She had high standards regarding the responsibility of good scientists, as well as in their integrity.

In her speech "Future of University Research" at the Centennial Celebration of the American University in Beirut in 1967, she emphasized that fundamental research in science must provide a certain freedom to the scientists and must be supported by the government.

She also emphasized that freedom and funding of the scientists must be judged solely by the merit of their work, and that "scientists must be sensitive to the needs of society, and must accept the responsibility coming along with freedom".[62]

Wu had profound insight into what constitutes a good scientist. She strongly believed that science education must provide the opportunity to develop creativity, and classroom work alone simply could not do it.

She said: "Students will not become great scientists by taking lectures, memorizing formulae, or doing routine experiments. It is more important to develop a habit of investigation, risk-taking, and the ability to observe and deduce."[63]

As science becomes a large part of modern culture, it must be accepted and appreciated by the general public. Wu believed that scientists must learn to communicate better, identify interesting topics, and be receptive to the public outside of the scientific community.[64] She had her own internal gauge for the integrity and discipline of scientists. In general, she believed that truly creative scientists would immerse themselves in research, and not waste any time in envy or on politics.[65]

In private, she did not care for many famous scientists. She once said that there were messy things going on at the upper echelons of scientists too, but those were kept from the public.[66]

Wu retired in 1980, and became a professor emeritus. Her office was still on the 13th floor, and she was often spotted in the library, or the auditorium in the 1980s.

Wu and Luke at the Wu Zong-Yi Memorial in his hometown, Liuhe, in Jiangsu Province.

Her office had two parallel long rooms. There were desks, chairs and bookcases in the room outside which used to be the laboratory. There was a simple sofa in addition to the desk, chair, and bookcase in the inside room. Two framed pictures hung on the wall. One was a picture of Albert Einstein, and the other her diploma of an honorary Sc.D. degree from Columbia University.

Einstein was the greatest scientist of the 20th century, and an icon of science. Those two pictures might have reflected Wu's assessment of science and herself. In his reminiscence of Madame Curie, Einstein said that the moral character, rather than the intellectual accomplishment, of a leader might have more impact in history, and intellectual accomplishment was more intimately related to personality. The assessment of Curie was equally fitting to Wu.

Wu's health deteriorated in the 1990s, and she became more inactive. She occasionally went to her office to take care of some correspondence, but mostly stayed in her apartment, five minutes away from the Pupin Laboratories. The apartment was on the seventh floor of an apartment building on Claremont Avenue, a property of Columbia University. The apartment had a combined sitting and dining room, as well as three bedrooms and bathrooms.

Wu in the office of Columbia University in 1974.

254

Part of the sitting room in the New York residence.

The Wu Chien-Shiung Foundation within Academia Sinica was established in 1992, as part of the celebration of her 80th birthday. With four Nobel laureates (*from right to left*), C. N. Yang, Samuel C. C. Ting, T. D. Lee, and Li Yuan-Zhe, and husband Luke (*second from right*).

◀

Wu in her New York City
residence.

Wu and Luke in Taipei in 1992.

▼

Wu had her 80th birthday in 1992. There were conferences and celebrations in Taiwan, China, and Europe in addition to Columbia University. She and Luke, born in the same year, both had high blood pressure and minor heart problems. In addition, she suffered from migraine headaches.

Wu endured a minor heart attack in middle of 1995. She recovered fairly well after having a pacemaker implanted, except for some effect on her eyesight. Wu and Luke lived independently and took care of themselves, sometimes eating out. They hired a nurse-caretaker after 1995. Except for receiving awards or attending conferences, they stayed home most of the time.

End Scene

New York City was still rather cold in March 1990. It was sunny and bright. Young students were scattered around on the stairs of the Low Library at Columbia University. The Low Library is a Renaissance-style building with tall stone pillars. It had long been converted to an administration building. Wu, wearing an overcoat, and Luke, overcoat and a hat, were struggling to walk up the long stairs to the administration building.

The students sitting on the stairs enjoyed their youth and vitality, and dreamed of a bright future. They did not pay much attention to this old Chinese couple. They could not have imagined that this short lady had experienced the same hope and dream of the future on the campus of UC Berkeley half-century before and ascended to the pinnacle of science with her brilliance and talent.

Wu and Luke reached the top, walked past the pillars, and went inside. The sunny world outside was particularly bright.

Notes

1. According to the handbook of the Nobel Foundation (1993–1994), the Nobel Prize in 1978 was worth 630,000 Swedish Kronor, or about 100,000 US dollars.
2. Interviews with Leon Lederman and Noemie Koller.
3. Letter to John Hasting, public relations department, Columbia University, from Wu, December 29, 1958.

4. Interview with Samuel Devons, March 16, 1990, office at Columbia University. Interview with Felix Boehm, October 20, 1989, Caltech.
5. Similar view from her collaborators Felix Boehm and Ernest Ambler, and her students Noemie Koller and Stanley Ruby.
6. Interview with Zhang Xi-Ying, August 15, 1989, London residence.
7. Interview with Luke Yuan, January 23, 1990, New York City residence.
8. Similar view from her collaborators S. Devons and Ernest Ambler, and her student Mo Wei.
9. Based on views of Zhang Xi-Ying, I. M. Pei, Adina Wiens, Margaret Lewis, J. Steinberger, and S. Devons.
10. Interview with I. M. Pei, November 29, 1989, office in Manhattan.
11. Interview with Jack Ullman, September 8, 1994, the Bronx.
12. Interview with Georgia Papaefthymion, March 13, 1990, office at MIT. ("Orgy" referred to a wild, free-for-all party.)
13. Interview with Noemie Koller, December 12, 1989, office at Rutgers University.
14. Interview with Wu Chien-Shiung, December 5, 1989, New York City residence.
15. Similar view from her students Noemie Koller, Georgia Papaefthymion, Evelyn Wu, Mo Wei, and Chen Shao-Zhong.
16. Interview with Wu Chien-Shiung, July 25, 1990, New York City residence.
17. Interview with Robert Serber, March 14, 1990, New York City residence.
18. Interview with Margaret Lewis, February 22, 1990, library at Harvard Observatory. Interview with Adina Wiens, October 16, 1989, San Francisco residence.
19. Similar view from George Volkoff, Zhang Jie-Qian, Ren Zhi-Gong, and Noemie Koller.
20. Letter to Luke Yuan from L. W. Alvarez, December 21, 1972. Provided by Luke.
21. Similar view from Norman Ramsey, Charles Townes, Maurice Goldhaber, Felix Boehm, and student Noemie Koller.
22. Interview with Luke Yuan, September 27, 1990, New York City residence.
23. Interview with Luke Yuan, September 8, 1989, New York City residence. Scientists who ever listened to Luke's playing the *nan hu* included V. Telegdi and Felix Boehm.
24. Interview with Norman Ramsey, February 22, 1990, office at Harvard University.
25. Interview with Wu Chien-Shiung, September 13, 1989, New York City residence.
26. Interview with Xu Jing-Yi, October 16, 1989, San Francisco residence.
27. Interview with Margaret Lewis, February 22, 1990, library at Harvard Observatory. Interview with Noemie Koller, December 12, 1989, office at Rutgers University.
28. Interview with George Volkoff, October 12, 1989, faculty club at the University of British Columbia. Interview with Wu Chien-Shiung, February 12, 1990, New York City residence.
29. Interview with Wu Chien-Shiung, February 12, 1990, New York City residence.
30. Interview with Wu Chien-Shiung, December 5, 1989, New York City residence.
31. Telephone interview (New York–Santa Fe, New Mexico) with Vincent Yuan, September 10, 1990.

32. Interview with Luke Yuan, February 23, 1990, Harvard Square, Boston.

33. Telephone interview (New York–Santa Fe, New Mexico) with Vincent Yuan, September 10, 1990.

34. Interview with Wu Chien-Shiung, July 31, 1990, New York City residence.

35. Interview with Wu Chien-Shiung, February 23, 1990, Lewis residence in Boston.

36. Revealed by Physicist Charles Browman, a colleague of Vincent at the Los Alamos National Laboratory, August 21, 1991, Sicily, Italy.

37. Interview with Mrs. I. I. Rabi, January 28, 1990, New York City residence.

38. Interview with Wu Chien-Shiung, February 12, 1990, New York City residence.

39. Same as 38.

40. Same as 38.

41. Interview with Wu Chien-Shiung, February 25, 1990, New York City residence.

42. Interview with Wu Chien-Shiung, August 21, 1990, New York City residence.

43. "Dr. Wu Chien-Shiung — The Chinese Madame Curie", Sun Duo-Ci, in *Report on the "Special Contribution Award" from the Chia Hsin Cultural Foundation*, July 1965.

44. Interview with Zhang Xi-Ying, August 15, 1989, London residence.

45. Letter from Wu to Xu Jing-Yi, December 21, 1976. Provided by Xu.

46. Letter from Wu to a friend, April 24, 1967. Provided by Xu Jing-Yi.

47. Same as 46.

48. Same as 46.

49. Same as 46.

50. Her classmate George Volkoff said that she could hardly speak English when she first arrived in San Francisco.

51. "Wu Chien-Shiung: The First Lady of Physics Research", *Smithsonian*, January 1971.

52. Same as 51.

53. Interview with Wu Chien-Shiung, September 25, 1990, New York City residence.

54. Same as 53.

55. Same as 53.

56. Same as 53.

57. Interview with Wu Chien-Shiung, July 5, 1992, *Reading Times*.

58. Interview with Xu Jing-Yi, October 16, 1989, San Francisco residence.

59. Interview with Noemie Koller, December 12, 1989, office at Rutgers University.

60. Interview with Wu Chien-Shiung, January 12, 1990, New York City residence.

61. Interview with Wu Chien-Shiung, September 13, 1989, New York City residence.

62. Speech at the Centennial Celebration of the American University, Beirut, Lebanon, Wu Chien-Shiung, April 19, 1967.

63. Same as 62.

64. Same as 62.

65. Written reply to questions from the media in Beirut, Wu Chien-Shiung.

66. Interview with Wu Chien-Shiung, January 2, 1990, New York City residence.

Name Index